Praise for
Qigong Meets Quantum Physics

Modern physics confirms what mystics have repeated throughout the ages: that ultimate reality is an integrated oneness, that all things are fundamentally connected. Bock-Möbius adds to the discussion by drawing parallels between quantum mechanics and the philosophical theories of qigong and traditional Chinese medicine.

Much more importantly, however, she advises and instructs the reader to go beyond the inherent limitations of the intellect and to directly experience the wholeness of creation through qigong practice. She welcomes the reader to the path of qigong, a path of experiential realization.

–Yang Yang, PhD, author of *Taijiquan: The Art of Nurturing*

This lucid book, light years beyond Capra's *Tao of Physics* (1975), is a must for modern cultivators seeking to reconcile scientific worldview, quantum theory, and *qi*-flow. Bock-Möbius goes far beyond philosophy, offering excellent qigong methods and expertly laying out Science vs. Qigong issues.

The book is a brilliant summary of physics in relation to Daoist cosmology as consciousness theory. Both ways embrace wholeness, with different experimental methods to overcome the illusion of separation via control of energetic forces. Peppered with fabulous translations from the Daoist classics, this book will kick the yin-yang of your Inner Sage into a new quantum orbit.

– Michael Winn, Healing Dao USA

Imke Bock-Möbius has produced an interesting meeting ground between the Western sciences and the neo-classical mystical practice of Chinese qigong. Her mastery of both science and qigong shows through, allowing the text to explore a nascent area of inquiry.

In the history of East Asian philosophy, this book represents the beginning of a new era where the models of science of the West will be used to explore the wisdom of the East. This book is a must read for students of qigong, as well as those of theoretical physics.

–Steve Jackowicz, PhD, Adelphi University

The experience of *qi* and non-usual states of consciousness in the practice of qigong inspire practitioners, teachers and so called masters to reflect on emerging new physics as a "pop" rationale for the underlying causes/ mechanisms of the qigong experience. I have written on this from the perspective of a doctor of Chinese Medicine, not thoroughly trained in physics, and get constant appreciation from readers.

Now, here is a book by a trained physicist who is actually conversant in the amazing details of the quantum realm. This book will become standard suggested reading for the Instructor Trainees of the Institute of Integral Qigong and Tai Chi (IIQTC). I am sure that anyone who wants to understand the experience of Qigong and Qi in contemporary terms will gain immensely from this very accessible book by Dr. Bock-Möbius.

–Dr Roger Jahnke, OMD, Director, Institute of Integral Qigong & Tai Chi

Qigong
Meets
Quantum
Physics

Experiencing Cosmic Oneness

Imke Bock-Möbius

Three Pines Press
P. O. Box 530416
St. Petersburg, FL 33747
www.threepinespress.com

Originally published as *Qigong Meets Quantenphysik*
© 2010 Windpferd Verlagsgesellschaft mbH, Oberstdorf (
www.windpferd.de) – Translated by Livia Kohn

9 8 7 6 5 4 3 2 1

Printed in the United States of America
This edition is printed on acid-free paper that meets the American Na-
tional Standard Institute Z39. 48 Standard.
Distributed in the United States by Three Pines Press.

Cover Art: Provided by Windpferd Verlag. Design by Brent Cochran.
Used by permission.

Library of Congress Cataloging-in-Publication Data

Bock-Möbius, Imke.
[Qigong meets Quanten Physik. English]
Qigong meets quantum physics: experiencing cosmic oneness / Imke
Bock-Möbius.
    p. cm.
Includes bibliographical references and index.
ISBN 978-1-931483-21-6 (alk. paper)
1. Quantum theory--Philosophy. 2. Qi gong. I. Title.
QC6.B596313 2012
530.12--dc23
                          2011044422

Contents

List of Illustrations

Fig. 1a: *Shiatsu Journal: Information der Gesellschaft für Shiatsu in Deutschland* 2008/53: 47.

Figs. 1b, 2a, 2b, 2c, 4, 6, 19, 21, 22 and qigong photos #1-#16: Imke Bock-Möbius

Fig. 3: *Zeitschrift für Qigong Yangsheng* 2004: 50-51. Uelzen: Medizinisch Literarische Verlagsgesellschaft.

Fig. 5: Willigis Jäger, *Suche nach dem Sinn des Lebens* (Petersberg: Via Nova, 2008).

Figs. 7, 8: Ken Wilber, *Up from Eden* (Wheaton: Quest Books, The Theosophical Publishing House 1981, 1996).

Fig. 9: Francisco Goya: *Der Schlaf der Vernunft gebiert Ungeheuer* (akg-images GmbH).

Figs. 10, 20: Windpferd Verlag Oberstdorf, 2010.

Figs. 11, 12: Jürgen Audretsch, *Verschränkte Welt. Faszination der Quanten* (Weinheim: Wiley-VCH Verlag, 2002).

Fig. 13: Wolfgang Demtröder, *Experimentalphysik 3: Atome, Moleküle, Festkörper* (Berlin: Springer, 2005) : 3rd ed., p. 68, fig. 2.88.

Fig. 14: *Das neue Fischerlexikon in Farbe* (Frankfurt: Fischer, 1981).

Fig. 15: Ernst Peter Fischer, *Die andere Bildung* (Munich: Ullstein, 2002).

Fig. 16: Fritjof Capra, *Das Tao der Physik.* (Frankfurt: Fischer, 1977).

Fig. 17: Lothar Schäfer, *Versteckte Wirklichkeit* (Stuttgart: Hirzel, 2004).

Fig. 18: Bernard d'Espagnat, in *Scientific American* 241 (1979), 158-81.

All used by permission.

Acknowledgments

I would like to extend my gratitude to all my predecessors: those who have inspired my work, on whose shoulders I stand, and in whose writings and sayings, as Zhuangzi notes, "many aspects were addressed earlier" (ch. 27). I would also like to thank the many teachers, students, and companions that I have had along the way, as well as all my patient and helpful friends. Most of all, I would like to acknowledge the support and understanding of my family and especially my dear husband Karl-Heinz, who also created the diagrams.

I would like to express special thanks to my teachers: Jiang Wuche 江武彻 (Fuzhou), Ren Zhuoling 任卓玲 (Beijing), Zhong Guoqiang 鍾國強 (Taipei), Jiao Guorui 焦国瑞 (Beijing), Ding Hongyu 丁宏余 (Nanjing), and Gisela Hildenbrand (Frankfurt a.M). Last but certainly not least, I would like to express my debt to my support team: Livia Kohn for the inspired translation into English and for publishing this book, Elaine Knörich and Simone Duxbury-Ziemer for careful proofreading, Eberhard Möbius for proofreading the translated version of the physics section, and Isolde Jäger for kindly taking the qigong photos.

Preface

How we classify the various things and events we encounter depends on the model or paradigm of reality, to which we subscribe. The more complex our experiences, the more likely we are to have open paradigms. In the process of digesting our experiences and relating them to each other, we often discover unexpected connections. This is what this book is about: the underlying principle that unites qigong, quantum physics, and mysticism—and quite possibly, other aspects of reality.

About twenty years ago, I spent six months in Beijing, just after completing my Ph.D. in physics. A budding young scientist, I wanted to get to know the world and share the results of my research. So I went to Beijing, sponsored by a cooperative venture between the German Max-Planck-Gesellschaft and the Chinese Academia Sinica.

As soon as I got there, however, I found that my Chinese colleagues worked with the same kind of equipment we used and that one of them had just spent eight months in Switzerland, working with one of the leading scientists in the field. The transfer of know-how, then, came about in a rather unexpected way. . .

Every spring Beijing is subject to massive sand storms that bring clouds of dust from the Gobi Desert. Smart people either stay indoors or, if they cannot avoid going out, wear face masks to protect their nasal and respiratory passages. Not so smart, I duly got sick, suffering from laryngitis and a sinus infection, with high fever and a great deal of misery. I underwent one ambulatory treatment with *qi* and was completely cured. That was quite a surprise! How was that possible? How had it happened?

Even before coming to China I had practiced taijiquan for a while, but I had no clear concept of how qigong and Chinese medicine worked. Inspired by my experience, I decided to extend my stay and learn acupuncture to get a better grip of these subjects. After my return to Germany, I followed this up with an intermediate academic exam in Chinese language and culture, as well as in philosophy, and then decided to pursue training in qigong through the Medical Society for Qigong Yangsheng.

Having undergone extensive qigong study and training, I eventually returned to my original fascination with physics, remembering the original impetus that had led me into science before my interest shifted into qigong. This shift, however, happened almost by itself. The connections between the two have intuitively always been clear to me, but writing them down in the form of a complete, integrated theory is something else again. This integration is the subject of this work. Please leave your preferred concepts of the world behind and follow me now.

1. Introduction

Human beings ask all sorts of different questions both of themselves and of life. The answers they find serve to provide their fundamental worldview. My own questions had to do with what operates behind visible reality: How is the universe built? Why am I here? What structure underlies that which we think of as our reality? Is there a universal principle that appears everywhere? How can I not only know but personally experience this?

These are hard questions. Since I first learned of qigong in a kind of spontaneous remission, I have been particularly interested in the answers it has to offer. At the same time, my perspective also goes beyond qigong into mysticism and quantum physics, thereby allowing an answer that is as encompassing and enlightening—and also as unprejudiced—as possible.

My basic position is that, if there is a fundamental principle in the universe, it must work behind and/or within *all* things. This matches the traditional Chinese vision as expressed, for example, in a prose poem by Su Zhe 蘇轍, the brother of the famous poet Su Dongpo 蘇東坡 (1037-1101). He says, "There is only one cosmic principle (*li* 理) among the myriad things. Their only difference is where they start." A bamboo painting by Wu Zhen 吳鎮 (ca. 1300) similarly contains the inscription, "Having fully realized the cosmic principle that rests in emptiness—what sorrows could still fill the heart?"

Beginning from this position, my reasoning is that, if this is the case, I can take various things at random and study them in more detail to discover the underlying pattern. Should I then find a commonality in all these things, I can come to certain conclusions that I can analyze further with regard to their wider applicability, expanding from the specific to the general following the principle of induction. The assumed ubiquitous principle, I suspect, has a lot to do with the concept and apperception of oneness inherent in the multiplicity of forms, respectively with the loss of oneness. This loss has initiated our persisting quest.

In qigong, oneness appears in various ways: it is in the roundness and completeness of each exercise; it is also present in the concept of Dao which I focus on in the first part of this work. Next I turn to mysticism, where oneness is experienced as union with the highest, the superior, the divine. From here I explore the natural sciences. They, too, know of oneness: with nature as, for example, in the manifestation of nonlocal phenomena, which are already being used in modern technology. My concluding synthesis, then, combines these various experiences of the underlying principle of cosmic oneness. To help the reader activate this understanding, I complete the presentation with a series of qigong exercises geared particularly to the topic.

The purpose of this book, then, goes beyond theoretical exploration and understanding. I hope to help alleviate and possibly even eliminate the sorrows that fill the heart through insights in and the experience of cosmic oneness. [1]

---

[1] Regarding Wu Zhen, see Pohl (2007, 41). On induction, see Schäfer (2004, 26). An introduction to the reconciliation of Daoism and modern physics can be found in Bock-Möbius (2009; 2011).

2. What Is Qigong?

Qigong as a technical term was coined in 1947. The key protagonist was the Communist Party cadre Liu Guizhen 劉贵珍 who, suffering from various serious ailments, searched for and found efficacious exercises among Daoist practitioners and completely cured himself. In the process, he discovered that there were multiple practices, often with widely divergent names, that had the same goal: to cultivate perfect *qi* through body and mind, breath and spirit. In 1956, he became the director of the qigong clinic in Beidaihe 北戴河 near Beijing. Since its treatments, especially for chronic diseases, were quite successful, qigong came to spread more widely.

Over the years, it has undergone various stages of popularity and repression, and today is one branch of Traditional Chinese Medicine (TCM) after acupuncture and herbal pharmacology. It combines physical, breathing, and meditation exercises into one integrated system, encouraging both still and moving practices. The practice of qigong is good for almost anything: to cultivate *qi*, to prevent ailments, to stimulate self-regulatory processes, to supplement other therapies, and to enhance rehabilitation. The Beijing physician Jiao Guorui 焦国瑞 (1923-1997), whose system of "nourishing life with Qigong" (*qigong yangsheng* 气功养生) I predominantly use, has formulated his key concepts in teaching poems. For example, "It is easy to practice *qi* but hard to regulate it." It is also not all that easy to tame, nourish, stabilize, and control *qi*.

The Chinese characters for "qigong" help to better understand the meaning of the term. This book uses the abbreviated form of Chinese writing, as common in mainland China, usually after presenting the name or term in transliteration (*pinyin* 拼音). However, here I would like to take recourse to the traditional form of the character for *qi* 氣: it also contains the word for "rice" (*mi* 米) underneath the abbreviated character 气, which is described as showing a sail blowing. The word as a whole thus indicates not only air, steam, gas, breath, and atmosphere but also "steam from cooking rice" and thus, by extension, creative pneuma and life energy. This is obvious in a way, since rice is the foundation of all Chinese nutrition.

The second word in the compound is *gong* 功, which consists of the two characters for "work" (*gong* 工) and "strength" (*li* 力). It means accordingly work, effort, merit, skill, success—all part of the fundamental tenet of "continuous practice." The term *qigong* could therefore be rendered "working with *qi*," which sounds a bit pushy. Thus, some people prefer "working on *qi*."

Chinese medicine operates from a different understanding of sickness and health than Western bio-medicine. Its exponents have less interest in bacteria and viruses and instead presuppose that the individual will be healthy as long as *qi* is sufficient and flows smoothly through the meridians, the polarities of yin 阴 and yang 阳 are in balance, and the five energetic phases (*wuxing* 五行) are in harmony. This state of health should be recovered and, once achieved, maintained and enhanced.

Jiao Guorui has provided six key points for successful practice. They include naturalness, the principle of "light above, firm below," and the close interconnectedness of stillness and movement. In addition, the point that "imagination and *qi* follow one another" is essential.[1] To stimulate *qi*-flow through the meridians and into each cell of the body, practitioners combine their movements with suitable visions and images.

For example, to perform the exercise "Push the Mountain," stand with firm roots, i.e., focus your mental images on your feet and allow your lower back to relax. Let the thumb and index fingers form an open triangle, thus exerting a light pressure on the "Mountain" as you push forward

---

[1] An excellent outline of the development and social dimensions of qigong is found in Palmer (2007). For Jiao's poems, see Jiao (1993a, #37-42). A scholar-practitioner who prefers the translation "working on *qi*" is Ute Engelhardt (1987, 16). Jiao's six points of practice are outlined in Jiao (1988b, 60-69).

from the center of the body. Use your imagination and begin to open (push) from the center, thus letting the *qi* flow from the body's core to its periphery, i.e., into your hands and feet. Then pull the arms back in again and conclude the motion, allowing the *qi* to come back and collect in your core. This single exercise accordingly activates the entire meridian system, notably the yin channels in the arms and the Bladder Meridian that runs across the head and along the entire back of the body.

The core or center from which all movements come and to which they return is known as the cinnabar or elixir field (*dantian* 丹田). The term goes back to operational alchemy, which was extensively practiced in the early centuries of the Common Era, and signifies the potential of the human body as *qi* to be refined and enhanced like metal in the alchemist's crucible.

The elixir field is an area below the navel where the *qi* collects and regenerates to once again flow into the organism. This process can be enhanced by using imagination to guide and direct the *qi* with the help of visual images. These images, too, arise in the elixir field and are projected from there into the body as if onto a three-dimensional screen.

All processes and events within the human organism in qigong are described in terms of Chinese medicine. The basic assumption is that bodily actions are managed through centers at the core of a network of energetic channels running through the body. Although they are often named after inner organs that are also familiar to Western bio-medicine, they are in fact energetic orbs in charge of extensive functions and qualities. Thus, for example, the *qi* contained in the Spleen orb and flowing through its related meridian is also responsible for keeping the blood in its veins and arteries.

By the same token, metabolism in Chinese medicine is always the transformation of *qi* in the orbs. Since qigong stabilizes internal balance and harmonizes the metabolic processes of the body, it has a positive effect on health. In accordance with this way of thinking, it is the task of the Spleen and Lung orbs—assisted by that of the Heart—to distribute the essential nourishment acquired from food, water, and air throughout the body. The Kidney and Small Intestine orbs, moreover, serve to filter the clear parts toward further usage while guiding the turbid parts toward elimination via the Bladder and Large Intestine.

All twelve orbs in the human organism are supplied with *qi* through the meridians and regulated with the help of acupuncture points. Qigong exercises activate the meridians and acupuncture points, not only to strengthen the individual's basic metabolic function, but also to enhance the person's inherent perfect *qi*: the combination of postnatal and primordial *qi* that keeps people alive. Postnatal *qi* is distilled from food, water, and air, as well as social and emotional contacts: it is stored in the Spleen orb. Primordial or prenatal *qi* is the *qi* supply received from Heaven and Earth through one's parents: it resides in the Kidney orb.

Qi provides all the basic energy needed for the various bodily processes of production and decomposition. In addition, it manifests in two further aspects: vital essence (*jing* 精), which is more concentrated and tangible and gives shape to external processes in the organism; and spirit (*shen* 神), which is subtler and provides the more psychological and spiritual dimensions. Together they are known as the Three Treasures: their presence is essential to maintain the life of the body and the person.

When speaking of qigong exercises for nourishing life, we generally refer to practices whose effect is directed inward and contributes to an overall strengthening of health. It is, of course, also possible to direct the fruits of one's *qi*-work outward: this happens, for example, in the martial arts and during presentations of apparently magical powers. However, even then, the foundation is the internal cultivation of *qi* and mastery of the body's energetic processes. In other words, whether one uses *qi* to split a brick or to stabilize health, is just a matter of how one directs it: in or out.

A long period of regular practice will lead to various positive developments: the meridians become more permeable, *qi*-flow is subtle, the metabolism functions smoothly, the joints are stronger and more flexible, the central and vegetative nervous systems are balanced, and the immune system is strong. People often note that they feel better and more vibrant overall, while noting a clear reduction in stress, both internal and external. That is to say, if you undertake qigong practice with perseverance and a bit of good fortune, you have embarked on a path toward an overall harmonization of health and life. It is a fascinating and engaging journey toward an increasingly clearer contact with *qi*.

It is important to avoid practicing immediately after a meal, under the hot sun, in strong wind, or under drafty conditions. Similarly, you should not activate *qi* when emotionally agitated. Never strive for fast re-sults. In all aspects of the practice, moderation is essential: doing every-thing in the right measure. [2]

---

[2] A brief summary of meridians and orbs in Chinese medicine appears in Bock-Möbius (1993, 25-33). For a good general survey, see Kaptchuk (2000). For details on the workings of the body in Chinese medicine in relation to Western science and Daoism, see Kohn (2005).

3. Wholeness in Qigong

And so I fell in love with qi.

What is qi?

Living life just as it was intended.

My first encounter with qigong was fascinating. Continuously coughing and blowing my nose, I kept disturbing the class I was taking upon the recommendation of my acupuncturist Ren Zhuo-ling 任卓玲 at the Xiyuan 西苑 Hospital in Beijing. She came along when I first went and offered to treat my heavy symptoms in between exercises with acupuncture. However, I had no inclination to try needles on my face or throat.

To help me out anyway, she called over a colleague who specialized in medical qigong. He hardly looked at me and seemed rather short and irritated, and I was no longer sure if I still wanted the treatment. However, I trusted my acupuncturist and reclined on the treatment table, closing my eyes. After a short phase of acupressure, I received my first treatment with *qi*—involving no touch whatsoever. After about twenty minutes, they told me to get up: I was a bit woozy and had no clear idea of how I was feeling.

The next morning I woke up free of symptoms. I was flabbergasted. Something significant had happened in that cold and unpleasant hospital room, filled with all those people. Something had not only made me feel way better, but had come to touch me in my innermost being. For several days I walked about as if on a cloud of *qi*. Then my rational mind started to kick in again and wanted to know what in the world had happened there. It wanted to sort, classify, analyze, learn, and understand—doing the kinds of things a rational mind tends to do.

This interchange between heart and mind has continued to serve me well: the heart checks if the path feels good and right for me, whether it is a "path with heart," containing the powerful challenge to go "all the way." The mind sorts and evaluates the resulting experiences. Once in a while, it needs to adjust my dominant worldview. Thereupon, whatever I have experienced and understood demands to find expression in words or movements. First and foremost, however, I stick to the words and ask: Why do we practice qigong?[1]

Why Practice?

J iao Guorui defines nourishing life through qigong as the "dialogue with one's own vital essence." We learn through practice how to recognize the dynamics of *qi*, and how to follow it while still maintaining the ability to control it. Qigong is a bridge in the cultural exchange between East and West. Containing key elements that help improve people's health, it serves as a powerful means for dealing with the current "increase in psychosomatic illnesses" due to information overload, cut-throat competition, and internal, emotional burdens.

Qigong is an art of movement and a method of self-cultivation. However, its main goal is not self-expression, but the realization of specific principles on both the physical and mental levels. "One can only be truly free if one knows the rules." If one starts without rules, "one will not be happy to take on any rules later on and will not achieve anything. Thus anyone who wants to learn needs to start by observing rules." This not only holds true for bamboo painting, as described by Li Kan 李衎 (ca. 1245-1320), but also for qigong. It is a simple method, easy to practice, yet infinitely difficult.[2]

---

[1] The expression "path with heart" goes back to Carlos Castaneda (1973, 13).

[2] On Jiao Guorui's teachings, see Jiao (1988a; 1988b); for more on the dynamics and control of *qi*, see Jiao (1993a, #1, #83). On importance in psychosomatics, see

The actual practice focuses on recovering and supporting the balance between yin and yang, as well as on purifying, moving, cultivating, and nourishing *qi*, thus fulfilling the fundamental conditions of good health in Traditional Chinese Medicine.

What, then, is *qi*? Already found on the bronze vessels and oracle bones of the Shang (16th-11th c. BCE) and Western Zhou dynasties (1122-771 BCE), it is best understood through practice. Wang Chong 王充 (127-200) describes *qi* as "the ultimate and unifying cause of all things." *Qi* is a term to express movement; and life itself is movement. *Qi* is both, material and spiritual, substantial and dynamic, as in general, "Chinese thought does not distinguish between matter and energy" (Kaptchuk, 1994, 46).

As *qi* assembles into grosser density, things arise, just like water condenses to ice; as *qi* disperses into subtler levels, they pass on, just like ice melts into water (*Zhuangzi*, ch. 22). All existence depends on *qi*— comparable to what Western culture knows as vital breath or pneuma.

The various exercises combined in qigong can be traced back in texts dating back over 2000 years. The earliest detailed descriptions of Chinese healing exercises (*daoyin* 導引) appear in manuscripts unearthed at Zhangjiashan 張家山 (186 BCE) and Mawangdui 馬王堆 (168 BCE). They are supplemented by the famous jade block inscription on *qi* circulation (4th c. BCE) and discussions in philosophical texts such as Lü Buwei's 呂不韋 *Lüshi chunqiu* 呂氏春秋 (Spring and Autumn Annals of Master Lü; 3rd c. BCE). Its underlying worldview largely follows Daoist philosophy and classical Chinese medicine.

Nourishing life through qigong means to pursue the art of living. The term "nourishing life" (*yangsheng* 养生) appears originally in the *Zhuangzi* 莊子, next to the *Daode jing* 道德經 the second major classic of ancient Daoism. Zhuangzi (365-290) lived around the same time as Plato (428-347 BCE) and Aristotle (384-322 BCE) in Greece, and shortly after the main Chinese sage Kongzi 孔子 or Confucius (550-479 BCE). Unlike these other thinkers, Zhuangzi finds the goal of human life not primarily in the recovery of social and political order through the ideal ruler. Instead, he pro-

Geißler (1996, 84). On the relation of qigong and rules, see Hildenbrand (2000, 6); Pohl (2007, 41).

poses the cultivation of the individual to the highest possible level of perception through a variety of practices.

For example, the practice of sitting in oblivion (*zuowang* 坐忘) allows the person to empty the mind and go beyond all distinctions between self and other, thus becoming one with Dao and being able to subtly perceive the true nature of things (ch. 6; see Kohn 2010).[3] "As the person acts in accordance with Dao, he or she is in harmony with Heaven and Earth" (Tan 2006, 84). Neo-Confucians under the Song (960-1279) integrated both Daoist and Buddhist concepts and came to see it as the central task of human beings in the world to purify their minds—"like finding a pearl in the depth of a muddy pond"—to thereby achieve moral perfection and wisdom (Ommerborn 2005b, 104).

Daoism as a philosophy is first and foremost an appreciation of nature, naturalness, and spontaneity. In addition, one of its key goals is to enable people to live in harmony with nature. This fundamental attitude is pervasive in qigong practice as, for example, in the overarching demand for naturalness, one of Jiao's key points. John Blofeld explains it using the metaphor of bamboo, "When the wind blows, the bamboo bends; when the wind calms, the bamboo stills—never, not even for a moment, considering the advantages or disadvantages of bending versus stillness."

Naturalness is a way of being in the world that is in complete "accordance with the laws of nature," merging body, mind, and spirit "without anything acquired or artificial." The natural course of things in Chinese is called *ziran* 自然, lit. "self-so," "spontaneous." To remain healthy, "people should always live and act in complete accordance with their own inner natural so-being."[4]

Disease arises due to disturbances of the inherent balance of the body and its *qi* circulation. The movement and dynamic of *qi* establish an inherent unity among the different body parts, as well as between humanity and the universe. Instead of directly attacking pathogenous factors, as

---

[3] On the history of *qi*, see Ommerborn (2004a, 73); (2004b, 105); Kaptchuk (2000). On *daoyin* and its early documentation, see Kohn (2008); Wilhelm (1948); Engelhardt (1987); Roth (1999).

[4] On naturalness, see Blofeld (1985, 214-15); Tan (2006, 89); Ommerborn (2004b, 101); Liu (1998).

Western medicine tends to do, Chinese medicine focuses on re-establishing order within the organism and preventing ailments through guidance to a moderate and healthy lifestyle.

Qigong practice, then, serves to develop, enhance, and regulate vitality within the individual. Vitality, as defined by the biophysicist F.A. Popp, is "the ability of the organism to establish a principle of order that cannot be set up with mere techniques." It is part of Dao even before creation, and its cultivation follows the laws of nature. When in doubt, we can find it through practical reason and inner intuition: there is no point practicing with force or under duress. When "things are not understood properly, all efforts of cultivation are in vain." Practice and proper understanding always have to go together.

The fundamental law of nature in this context is the ongoing mutual transformation of yin and yang. "*Qi* represents an energetic potential, activated by the polarities of yin and yang." It is important to activate and maintain this at a high level, ideally through persistent cultivation. In other words, we need to develop a certain level of practice skill (*gongfu* 功夫) to transform the various energy-consuming activities of daily life into life-giving, energetic processes and thereby prevent premature aging. The better our *gongfu*, the more we can achieve positive energetization, not only on an accidental, occasional basis, but through purposeful, intentional action.

By practicing qigong in body, breath, and spirit, we stimulate our spontaneous healing forces: we mobilize and enhance latent body energies to prevent diseases, maintain health, support therapies, and find creative forms of self-expression. Whatever state we may be in, it is always advantageous to nurture our vital forces. Moreover, there are practically no contraindications, the exception being serious psychotic states that require special care in the use of visualizations and mental images. [5] Physical limitations are no hindrance to practice, since qigong can be undertaken with equal efficiency while standing, sitting, or lying down.

---

[5] On disease and practice, see Despeux (2007, 27); Zumfelde-Hüneburg (1994b, 67); Ommerborn (2003, 56); Neuhaus (2001, 107).

The practice of qigong activates yin-yang transformation in a variety of ways: movements and mental attitudes of opening and closing, body positions of high and low, motions of rising and descending, bends and stretches of the body, storage and activation of *qi*, fullness and emptiness, exchange of inhalation and exhalation, tensing and relaxing, movement and stillness. Transitions are always smooth and flowing; physical movements are the foundation of practice; but the mental intention is the guiding force.

How important consciousness is in living properly is well documented in the *Xinmulun* 心目论 (On Mind and Eyes) by Wu Yun 吳筠 (d. 778): ideally, we should live well-centered, without being influenced by external things, without letting our heart and mind be pulled into the external world by various sensory impressions, and without being separated from Dao. In doing so, we would have fewer worries, and it would be much easier to achieve the ideal state of sagely wisdom. To achieve this, however, the senses have to cooperate: especially the eyes have to serve the heart and mind, contributing to the calming and stabilization of consciousness.

"Nourishing life with qigong is a separate science of human life processes; it focuses on nourishing the vital force." Jiao Guorui divides this science into nine areas, including the origins and conceptions of qigong, classical literature and theoretical foundations, related fields and practical applications, as well as experimental research.

Independent of which area of life we want to benefit from *qi* cultivation—healing, health maintenance, artistic expression, or personal unfolding—in all cases, the key is the positive attitude and energetic well-being in a continuous flow-state that we reach through practice. What, then, is energetic well-being? Before I discuss this in some detail, let me briefly explain the basics of Daoism, which forms the foundation of qigong and which Needham calls "the root of Chinese culture."[6]

---

[6] On the mental dimensions of qigong, see Hildenbrand (1993, 6). For a translation and discussion of Wu Yun's work, see Kohn (1998). For Jiao's nine areas, see Jiao (1993a, #9); (1992, 262); Geißler (1996, 83).

Daoist Visions of Oneness

A lready the *Shiji* 史记 (Historical Records; 104 BCE) notes that Daoism took the idea of ordered processes in nature from the yin-yang cosmologists. The term *dao* 道 that gave the school its name goes back to the classics *Daode jing* and *Zhuangzi*. The character consists of the word for "head" on the right, combined with "foot" on the left. Dao is thus the way, the flow of things, the order of the world, or the fundamental principle of all things. It indicates an underlying structure that is immanent in all things, can be perceived intuitively, and pervades the cosmos like the threads of a net.

As and when people follow the course of Dao, everything flows in its proper way. To achieve this, they should adopt an attitude of nonaction (*wuwei* 無为), which means "abstaining from meddling with the spontaneous flow of things." The original meaning of the word *dao* as "way," moreover, may have to do with the fact that people's living conditions in the old days, in terms of such aspects as the level of social order and potential dangers of raids and violence, depended to a great deal on whether the roads in their state and country were passable or blocked. However, we can speak of Dao time and again, yet its essence remains elusive. As the *Daode jing* says:

> The Dao that can be Dao'ed is not the eternal Dao.
>
> The name that can be named is not the eternal name.
>
> The nameless is the origin of heaven and earth;
>
> The named is the mother of the myriad beings. (ch. 1)[7]

The language of logic and reason is useful in everyday life, but fails in matters of its underlying, cosmic laws. For this reason, the ancient Chinese developed a variety of modes to describe the origin and development of the universe. Wu Yun, for example, speaks of the Nonultimate (Wuji 無

---

[7] For a discussion of nonaction, see Ommerborn (2004b, 113); Liu 2001. On the original meaning of *dao*, see Ommerborn (2005a, 42). For the *Daode jing*, see Schwarz (1985). An easily accessible translation of the *Daode jing* can be found at http://ctext.org/dao-de-jing.

极), a state of empty chaos before creation, from which "existence emerged and developed into Dao."

"Dao materializes as *qi*," says the internal alchemy classic *Xingming guizhi* 性命圭旨 (Superior Pointers to Inner Nature and Destiny; dat. 1615), which integrates the three schools of thought of Daoism, Confucianism, and Buddhism. Here Dao is the original ground from which cosmos and humanity emerge.

Jiao Guorui, too, notes that creation began with the Nonultimate, from which the Great Ultimate (Taiji 太极) arose. "Taiji represents the original state before the separation of yin and yang," i. e., the one. "The two forms yin and yang develop through inherent movement;" they represent the sunny and shady sides of a mountain, the lighter and denser energetic aspects of all things. Wang Zongyue's 土宗岳 (18th c.) still frequently used *Taijitu* 太极图 (Symbol of the Great Ultimate) clearly shows the dynamics of the two polarities: each side embeds the seed of the opposite pole and thus contains the potentiality of transformation.[8]

Since the opposites are inseparable and interdependent, they function as value-neutral, natural principles. Only in the writings of the Confucian thinker Dong Zhongshu 董仲舒 (179-104 BCE) is there a description of yin as morally negative and yang as positive. As yin and yang are joined in oneness, we call them complementary or speak of them as polarities.

Daoist thinkers after the Han, moreover, regarded humanity as a replica of the cosmos and decided that it was subject to the exact same laws. Thus the Taiji symbol serves as the most compact depiction of our life in the world of opposites (see Fig. 1a). It teaches us that day and night are both necessary to make up the twenty-four hours of a day, that life and death are both essential for the wholeness of existence, and so on.

On the level of concrete phenomena, Dao or Taiji manifests itself in various forms. However, each form completely contains the one Dao—in the same manner that lakes at night reflect the one moon.

---

[8] On the integration of the three schools of thought, see Ommerborn (2007a, 70); Darga (2004a, 64; 1999). For more on the Taiji symbol, see Jiao (1996b, 7); Louis (2003); Ommerborn (2006, 20, 31, 34).

Fig. 1a: The Taiji symbol

Dao in this metaphor is equivalent to the moon and the various re-flections are analogous to the things of the world. As human beings, we suffer as a result of the separation from original oneness and it is our over-arching goal to fully dive into the flow of life, to recover our true identity as part of Dao and to return to it (*Daode jing* 25, 40, 42).

This can be accomplished through a variety of methods. One is the Daoist meditation known as "guarding the One," a form of concentrative immersion, through which one may recover "what has been lost from the One." Another is to "cultivate *qi* to return to the origin;" yet another is to achieve one's "true inner nature" (*Daode jing* 10). Realizing oneness above and beyond all things and "transforming this realization into concrete thinking and acting in the world" are the main characteristics of the sage. In general, deviating from Dao leads to degeneration (as obvious, for ex-ample, in the current state of society); recovering naturalness opens the way back to Dao. The symbol of this return is the circle: Dao is depicted as an empty circle (Fig. 1b).[9]

---

[9] On "guarding the One," see Kohn (1989a, 127). On the Daoist sage, see Ommerborn (2004b, 108); (2005a, 43). For the empty circle calligraphy, see *Shiatsu Journal* (2008, 47).

Fig.1b: The empty circle

There are many modes to describe cosmic unfolding, identifying the beginning as emptiness, nonbeing, *hundun* 混沌, chaos, or absolute quietude. Whichever it may be, in all cases there is a fundamental principle that is different from empirical things and can exist independently—the state "before Heaven" or of precreation (see Fig. 2a).

Nonultimate

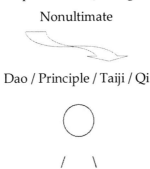

Dao / Principle / Taiji / Qi

Fig. 2a: From nonbeing to being: the development of the cosmos

Dao contains everything: it brings forth things and is the place to which they all return. It is the immanent oneness of things. Neo-Confucian thinkers identify it with the term "principle" (*li* 理). Wang Zongyue notes, "Taiji arises from Wuji," and Zhuangzi says, "Dao precedes Taiji" (ch. 6).

The *Xici* 系辞 (Great Commentary), a Han-dynasty supplement to the *Yijing* 易经 (Book of Changes) similarly insists, "Taiji arises from the changes; it brings forth the two forms [yin and yang]" (Fig. 2b). The *Daode jing*, on the other hand, says:

> Dao produces the One;
>
> The One produces the two;
>
> The two produce the three;
>
> And the three produce the myriad beings. (ch. 42)

Dao and *qi* were seen as identical as early as the Zhou dynasty, and the Neo-Confucian thinker Zhu Xi 朱熹 (1130-1200) similarly says, "In the universe there is only one *qi*; it divides into yin [and] yang" (see Fig. 2b).

Dao / Principle / Taiji / Qi

Fig. 2b: The development of the polar forces from original oneness

Zhou Dunyi's 周敦頤 (1017-1073) *Taiji tushuo* 太極圖說 (The Symbol of the Great Ultimate Explained) notes that the ongoing mutual transformation of yin and yang brings forth the five phases, which in turn generate the myriad things (*wanwu* 万物).[10]

The introduction to the *Yijing* describes the immutable law that pervades all change and transformation as the underlying sense, the one in the many. "To realize oneself, there needs to be . . . a positioning. This fundamental status leads to the great beginning . . . : the Great Ultimate." Before the elementary positioning happened, there was the Nonultimate,

---

[10] For Taiji in the *Yijing*, see Wilhelm (1950, no. 48). For Zhu Xi on *qi*, see Ommerborn (2005b, 102). For a translation of Zhou Dunyi's work, see Chan (1963).

shown as the empty circle. The Great Ultimate, the universe on the verge of creation, on the other hand, is the circle divided into yin and yang.

These multiple and contradictory outlines of how the world came into being are rather confusing. For this reason, I propose to pull them together into an integrated presentation that acknowledges all essential aspects and provides a meaningful structure. I see the unfolding of creation as occurring in five stages (see Fig. 2c): the first three match the ultimate beginning in the *Zhuangzi* (ch. 2), the last two reflect Zhou Dunyi's *Taiji tushuo.*

Step 1: The universes rests in the nonultimate or a state of nonbeing (*Daode jing* 40), a condition that exists independently of all empirical things.

Step 2: Dao forms the original ground from which all things arise; an empty circle, it symbolizes the immanent oneness of things.

Step 3: *Qi*, the universal power of change, begins to move and Dao divides into the two forces yin and yang, shown in the Taiji symbol as still connected into oneness yet already differentiated.

Step 4: Yin and yang transform in mutual interchange and the five phases develop.

Step 5: The five phases give rise to the myriad things, i.e., everything within us and in the outside world.

The *Daode jing* is the first to describe the transformations of Dao. It has continued to dominate Chinese thinking and culture over the millennia and was the subject of hundreds of scholastic and practical interpretations that each "distilled its specific visions of the world" on its basis.

The book outlines a unified world in which each and every part is connected and related to another. [11]

As recorded in the *Shiji*, the text goes back to the ancient thinker Laozi 老子 (ca. 500 BCE). According to the legend, he served as an archiviste at the Zhou court and instructed Confucius in ritual matters.

---

[11] For a collection of articles on the various dimensions of the *Daode jing* in history, see Kohn (1998); Robinet (1989).

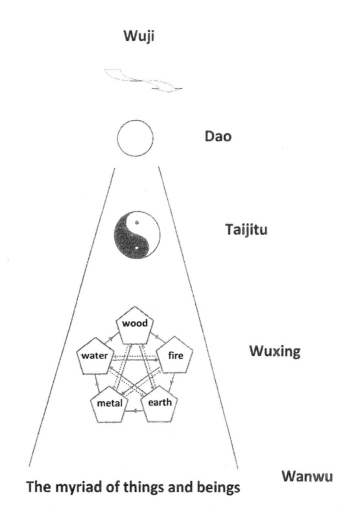

Wuji

Dao

Taijitu

wood

water   fire

Wuxing

metal   earth

The myriad of things and beings   Wanwu

Fig. 2c: The development of the myriad things through the five phases

Deploring the increasing moral decline of the Zhou kingdom, he decides to emigrate to the west. Upon reaching the western border crossing, he is stopped by the customs officer Yin Xi 尹喜 who recognizes the sage in him and urges him to write down his teachings, i.e., the *Daode jing*. Many stories about Laozi are fictional and it is quite possible that there never was a single master who encompassed the entire wisdom of Dao. It is more likely that the text developed as a compilation of various sayings and Dao-inspired schools over several centuries (see LaFargue 1994).

This does not diminish either the philosophical relevance or the cultural importance of the work. Its core concept of Dao, the ineffable power at the root of creation, can only be perceived by intuition. It contains the world in its original state, while also unfolding and developing it. In daily life it appears as the uninterrupted transformation of the complementary poles yin and yang. This means to escape harm and reach fulfillment in this world, one must adapt to the rhythms of these forces.[12]

As the cosmos unfolded over time, human consciousness gradually separated from Dao. Social and cultural structures increased in complexity, leading to an overall sense of disorder. In response, the Confucians believed that one should consciously adapt and nurture certain virtues to recover and maintain order—virtues that had been natural to the people of prehistory.

Daoists, on the other hand, subscribed to the understanding that the mere return to simplicity and mental stillness would recover the goodness in all things. This would lead to an overall state of perfect nonaction, of letting the things of the world be and unfold by themselves, of spontaneity (*ziran*). The sage who fully lives this ideal, moreover, is the central figure in the realization of complete harmony on earth. He stands in relation only to Dao, beyond the mutuality of ordinary relationships, and seeing one's aloneness is accordingly a "key concept in Daoist mysticism."[13]

To act in accordance with Dao is nonaction; always remaining in nonaction so that "there is nothing that is not done": that is Dao (*Daode jing* 37). Daoist thinkers accordingly emphasize withdrawal from ordinary society, insight into one's natural inner impulses, and a strong focus on the individual.

---

[12] For an unraveling of the legends surrounding Laozi, see Graham (1990); also reprinted in Kohn and LaFargue (1998). For details on the Daoist understanding of yin and yang, see Graham (1986); Kohn (1997).

[13] On spontaneity, see Ommerborn (2004b, 102). For a discussion of "seeing one's aloneness," see Kohn (1992; 2010, 22).

Both Daoism and Confucianism have a cosmological orientation. Dao supports the world and everything into becoming, then moves them forward toward passing on. Harm only occurs when there is disharmony with the order of Dao. Characterized by rhythmic changes and transformations, Dao is also a model for human behavior, social order, and cultural organization. Yin-yang and the five phases effectively describe the ongoing processes in the cosmos as well as in state, society, family, body, and so on: there is a pervasive correspondence between macrocosm and microcosm.

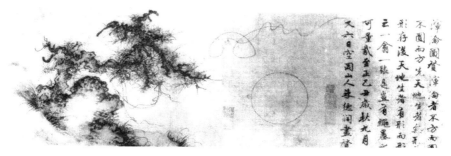

Fig. 3: Zhu Derun (1294-1365), "Chaos" (Hunlun). Shanghai Museum[14]

Everything has its place in this synchronistic order, and if applied correctly, everyone can return to cosmic harmony. "The movement of Dao is return" (*Daode jing* 40; see Fig. 3).

This transformation of return depends on the merging of complementary pairs. The intuitive or individual realization that one is originally one with Dao, i.e., the actualization of universal Dao within the individual person, is called "De" 德, which means virtue or inherent potency.

Zou Yan 騶衍 (ca. 350-270 BCE) was the first to formally expound the theory of the five phases as a fundamental pattern of classifying the myr-

---

[14] The powerful landscape elements pine and rock in the left part of the painting symbolize the temporal duration of all matter. In contrast, the threads floating in empty space on the right side show utter transitoriness. Duration and transitoriness form different aspects of a larger reality and form one Dao (the empty circle). Similarly, painting and words join in a larger whole, see Goepper (2004, 50).

iad things and of understanding dynamic developments.[15] He also presented a stronger appreciation of the concept of *qi*, clearly defining it as the core matter or root of the world, the original power of Dao and of all things, however minute. In his system, yin-yang and the five phases are concepts to define the strength and quality of *qi*. *Qi* is the bridge between the visible and invisible worlds, between all things and Dao.

From a religious perspective, the question arises in this context, how what we generally call evil arises in the world. On the one hand, it is inherently part of nature (yin-yang): as soon as things come into being, they are defined in terms of polarities, which means that the definition of good creates the presence of evil. As the *Daode jing* says:

All in the world know the beauty of the beautiful

And in doing so gain an idea of what ugliness is.

They all know the skill of the skillful

And in doing so know what lack of skill is.

Thus being and nonbeing bring forth each other,

Hard and easy complete each other. (ch. 2)

In other words, evil is necessary to create the good. Already Goethe in his *Faust* has Mephisto say: "I am part of that power, not understood, which always wills the bad, and always works the good."[16] The *Zhuangzi* similarly notes, "Once the transformations through virtuous heroes are done away with . . ., life in the world can finally be in line with the supernatural" (ch. 10). All this is part of natural evil and is inherent in life: we have to accommodate its presence.

On the other hand, there is also evil understood as the "result of human disturbance of the cosmic rhythms." Created evil that can be avoided or attacked, arises "through the intentional breach of cosmic harmony;" it was later mythologized in the form of demons in religious Daoism. "Each

---

[15] On virtue as inherent potency, see Ommerborn (2005a, 45); Kohn (2010). For more on Zou Yan, see Chan (1963); Gernet (1988, 93).

[16] Passages from Goethe's *Faust* are taken from the edition in Goethe (1963). An easily accessible edition is found at www.gutenberg.org. For an edition of *Zhuangzi*, see http://ctext.org/zhuangzi.

individual has the task to clearly recognize and control his or her internal tendencies." Human beings have to work constantly to recover harmony—both in daily life and in the practice of qigong. If we lose sight of this task, we may quite possibly enter a state of chaos. As Zhu Xi describes the cosmic cycles, "There comes a time when humanity completely loses Dao and all returns to the state of primordial chaos (*hundun*)." Everything perishes, and then creation begins anew.[17]

Practicing with the *Taiji* Symbol

All of these concepts are variously activated in practice. Thus the Taiji symbol (see above, Fig. 1a) serves as an aid to meditation, and the very character for "ultimate" (*ji* 极/極) shows "a person standing between Heaven and Earth like a tree." As reflected in the qigong exercise of "standing like a tree," this is a tree that reaches beyond; it points to something far greater than ourselves (see ch. 7 below).

The circular Taiji symbol is an inspiration for practice. It challenges us to contemplate wholeness both during practice as well as in our personal lives: in each exercise, the polarities should merge into one another easily and smoothly like day and night. We induce their rhythmic balance through variations of opening and closing, rising and descending—the four main movements of *qi*. The most concise description of *qi* practice is "opening and closing," especially since rising and descending are just the vertical variation of this as seen from the center.

Each qigong exercise is whole and complete. If we eliminate opening, there is no place to do any closing. "When yin and yang separate, the essence of life is exhausted." However, it is particularly impressive that completeness continues throughout. Even as we practice opening, a certain level of closing remains present to prevent excessive motion: opening thus always also contains the seed of closing. Both in concentrated and controlled widening, as well as in relaxation, we maintain stability. By the

---

[17] On human control of internal tendencies, see Kohn (1997, 99). For more on "hundun," see Ommerborn (2007, 90). Renewal as a Daoist theme appears in Girardot (2009).

same token, each closing contains a seed of opening to prevent tightness. We retain a certain width to allow *qi* enough extent, and each forceful position still entails flexibility.

This means that each exercise combines the opposites in mutual harmony, guided most of all by conscious intention, a fact that needs to be kept in mind in the rendition of "qigong" as "Chinese breathing practice." Each exercise only comes into its own through conscious intention—just working with its outer aspect misses the most important part and renders it empty. This in turn means that "whichever method best opens the right internal state is the most effective."[18]

From a different angle, looking at qigong as an art of living, the Taiji symbol also contains guidance for our personal lives. How can we integrate the different poles in our daily life? Can we see Dao in anything, whether we like it or not? Have we ever tried to see joy and effort, pursuit and fulfillment as complements rather than opposites? Have we experimented with bringing them into harmony as an experience of wholeness, allowing one to shine forth behind the other? Instead of insisting on opposition, this helps us to create constructive relationships among things or events. Another major prerequisite for a good mental practice state, besides wholeness, naturalness, and mental focus is something called "stepping into stillness" (*rujing* 入静).

Stepping into Stillness

To correctly practice qigong, it is essential to find the right state of mental stillness—simultaneously focused and relaxed. It is a core condition to ensure the regenerating effect of the practice. Although one is deeply relaxed in this state, it is neither like sleep nor ordinary rest. The key to understanding this state is the perception that stillness and movement are complementary polarities that transform into one another.

---

[18] Moving with the Taiji in qigong practice is the subject of Hildenbrand (1993, 7); 2007; Jiao (1993a, #43-51); (1988b, 52). Adapting the practice to one's inner state is a key point in Jiao (1996c, 36); Hildenbrand (2001, 9).

As the Taiji symbol shows, stillness at all times also contains movement, while all movement always also contains stillness. Both together are Dao.

The result of this is not, as Jiao Guorui outlines, a practice in complete oblivion or unconsciousness, not knowing "which method we use or what conditions need to be met." Rather, we practice in subtle clarity, a level of wakefulness that allows us "to feel gentle internal shifts and movements of life." We learn to screen out disturbing impulses and reach a brain state of "relative inhibition," which clearly shows on EEG monitors.

This creates ideal conditions for self-regulating and regenerating processes in the bodymind, such as, for example, the subtle balancing of physical functions. The mental state is sometimes expressed in metaphors, such as a mirror or water: the mind is like the calm surface of a lake, stirred occasionally through soft breezes (sensory impressions) whose effects pass by quickly (thoughts moving on). It is like a clear mirror, which only reflects our face as long as we look at it, remaining otherwise serene and undisturbed.

The Daoist thinker Wang Bi 王弼 (226-249) describes this state of mind by saying that one may react to and deal with outside things and events but does not get entangled in them. He connects this to the *Zhuangzi*, which says about the emotions that ideally "one neither pursues things nor comes to meet them: one mirrors them without holding on to them" (ch. 7).

Stillness is understood as a special state of movement: it is "movement in stillness" and does not mean being motionless. Physicists overcome this issue by using the variable "v" in their equations to represent velocity instead of using specific numeric values. This means that the numeric value is of no importance and one can express stillness as motion at "$v = 0$"—the central aspect of movement remains. By the same token, we understand stillness in qigong as a special kind of "force": it is the emptiness of the mind.[19]

---

[19] Descriptions of stillness in practice appear in Cobos-Schlicht (1998, 21); Jiao (1996a, 196). Its physical manifestation in brain waves is discussed in Bock-Möbius 1996; Haffelder (2006, ch. 3); Jiao (1993b, 12, 13). On Wang Bi and his line of thought, see Ommerborn (2004a, 75). Movement as part of stillness is emphasized in Jiao (1993a, #56).

How, then, do we enter this state of stillness? To prepare, we take good care to place body and mind in a framework of naturalness by matching our activities to the respective biological cycles and learning to be overall generous, optimistic, and calm, without applying any artificial means for health maintenance. In actuality, this means we establish a regular daily practice for ourselves and our students: this eases the transition from the turbulences of everyday awareness to stillness.

The transition itself is best effected by following a specific sequence of slow, gentle movements, taking on a particular position of stillness, breathing deeply and consciously, or by visualizing a particular image that induces a state of relaxation. One may also focus one's entire intention in the elixir field, follow the breath with one's attention, or vocalize a certain sound or chant: any of these methods will collect and still one's thoughts—often compared to a horde of wild monkeys or galloping horses. Initially quite difficult, this becomes easier as the mind habituates to the practice, adopting a new reflex toward stillness. In all cases, however, it is important to maintain a positive state of mind from the beginning—its enhancement also being the result of the practice.

Since *qi* is at the root of creation and of the continuous unfolding of all things, everything in the universe—including humanity—consists of *qi*. Humanity and nature are closely related. We can experience this relationship in our own bodies, but only if we practice stepping into stillness. Only in a state of inner quietude can we know what it means "to match nature," as Liu Tianjun 劉天君 points out (2005, 90-91). This means that stepping into stillness (*rujing*) and being natural as part of the cosmic self-so (*ziran*) form a feedback system: they depend on and determine one another.

The mental state of stillness is also the goal of Buddhist meditation. It is known in this context as "neither think nor not think" and is "the mental core of enlightenment," which in the Chinese (Mahayana) context means "the realization of identity with the absolute." In other words, it indicates the "boundless experience of complete universal oneness, where the duality of the world is overcome." Jiao Guorui's tenth point of practice similarly speaks of "concentrating the mind and transforming it toward emptiness . . ., cultivating emptiness and attaining oneness with Dao" (1997b, 77).[20]

To sum up, I would like to describe this mental stillness as a state that goes beyond all polarities and affords us a first glimpse of wholeness, oneness, and universal connectedness in a whiff of Dao.

The Flow State

Another major condition for attaining the positive effect of qigong is the open flow of inspiration based on a proper understanding of the internal conditions to be met in every exercise. Inspiration grows from suitable visualizations of natural phenomena, landscapes, or poetic metaphors. This is defined as the "merging of outside things with one's own feelings." As we enter the proper state of practice, the visualizations become tangible and we can feel just how the various movements of opening and closing come together to create one integrated whole.

Our postures and movements activate the fundamental principle of change and transformation between yin and yang. As and when the movements become smooth and harmonious, Dao begins to "flow through" us, as Blofeld puts it (1985, 39). We strive for wholeness in each and every move, in each and every exercise - a wholeness that is above and beyond the polarities is Dao. However, our very own body is nothing but a replica of Dao. Just as yin and yang are inseparable and form the Great Ultimate, the One, so is every single being a Great Ultimate, a cosmos in miniature that follows the very same rules as the universe at large. All this is first felt in the proper state of practice.

In terms of personal awareness in this state, we feel healthy and vigorous, qi flows freely, and there is a sense of timelessness. This state of "complete well-being in body and mind reflects a key ideal of Daoism." It is a terrific level of overall integration, a feeling of being part of a greater

---

[20] On the mind as monkey, see Jiao (1993a, #58); (1992, 201). For a positive mental state during practice, see Boente (1999, 92); Jämlich (2000, 77). The boundlessness of the experience of oneness is discussed in Cobos-Schlicht (1997, 19).

whole that lends meaning to our existence. In addition, the more we feel connected to the greater cosmos, the easier it becomes to use its powers.

An example of this appears in Su Dongpo's poem on the painter Wen Tong 文同. It shows how the love of painting and of Dao inspires the artist to the point that he becomes the bamboo that he paints, eliminating the separation between subject and object, humanity and nature. This is the key to his great mastery. The same holds true for the bell stand maker in the *Zhuangzi* who fasts for seven days before even cutting the wood and produces a work of art people consider divine (ch. 19). Meister Eckhart (1260-1328) says in similar terms, "When a master creates an image . . , he does not take the image into the wood but instead cuts off all outside wood that conceals the image contained in its depth."

A major tool for increased awareness of the „concrete oneness of our body with the world surrounding us" is the breath.[21] Already the *Zhuangzi* notes that the perfected people of old "breathed all the way to their heels," while ordinary people only „breathed with their chests" (ch. 6). Imitating this in qigong, we use every single in-breath to infuse the pure *qi* of the cosmos into ourselves, while actively removing stale *qi* from the body with every out-breath. In addition, by feeling increasingly better, we are equipped to face the darker sides of ourselves and allow them to be trans- formed into wholeness, thus enhancing our personal integration and well- being.

The feeling "that body and mind are separate is very unpleasant," a practitioner notes after only "feeding his head" for many years. We also know this from literature, most notably Goethe's *Faust*, "Two souls, alas! reside within my breast, and each withdraws from, and repels, its brother"(see www.gutenberg.org). Qigong practice allows us to overcome this feeling, and many people report a sense of "peace and inner align- ment," often also described as a deep stillness that was inaccessible before.

---

[21] On inspiration, see Jiao (1996c, 37); on wholeness in the exercise, see Hil- denbrand (1996, 14); on the cosmic rules, see Jiao (1997a, 10). For the Daoist state of well-being, see Kohn (1989a). Wen Tong and the key to great mastery are de- scribed in Pohl (2007, 36). On how a master creates an image, according to Meister Eckhart, see Quint (2007, 144). The breath as a tool for awareness is exemplified in Schild (1997, 59).

Others note how much their overall attitude to life has improved since they began their practice and how much deeper their breathing becomes during practice.

Some note that their bodily movement structures have changed through practice, making it easier to transform underlying patterns and improving self-esteem. The senses are more acute and the mind is more relaxed and at ease. Even a short period of practice makes practitioners feel more relaxed, balanced, and energized. Full of an increased sense of well-being, they maintain internal stillness and a positive attitude to life. More specifically, the practice alleviates depression, coronary heart disease, and high blood pressure, as well as asthma and migraines.[22]

Qigong practice also helps maintain a good balance of body, mind, and spirit. It strengthens our ability to react appropriately to difficult situations, thus reducing stress and enhancing health and vitality. Through the practice of qigong we take on the responsibility for our body, coming to "feel vibrantly alive all the way to our fingertips."[23] On another level, qigong makes it possible to influence psychological processes, without completely replacing psychotherapy, by affording deeper insights and allowing the personality to mature. In this respect the practice matches ancient Chinese wisdom: already Confucius says that the path of the disciples is an ongoing maturation, not just the acquisition of knowledge.

Qigong has holistic effects: however, rather than working to cure a specific condition, it dissolves the underlying conditions that lead to ailments. Even practice systems geared toward the regulation of specific functions have an impact on the entire person. Yet it is not possible to completely replace one series with another; the practice as a whole remains just one factor among many that lead to good health. Activating

---

[22] See Hofmann (1999, 105-06), Ritter (2000, 82-87), Reuther (1996, 44-50), and Friedrichs (2003, 101-12) for studies of the impact on qigong on these various ailments.

[23] On pleasant and unpleasant feelings during practice, such as body and mind as "separate," or on improving self-esteem, see Jiao (1994, 7-8); on peace and inner alignment, see Reerink (1997, 96). The behavior of the breath during practice is demonstrated in Zumfelde-Hüneburg (1994a, 15). On a positive attitude to life, see Sandleben (1997, 113); on feeling "vibrantly alive," see Boente (1999, 92).

deep-seated personal resources through qigong emphasizes inherent powers of the individual, "enhancing what is proper" and "eliminating the
pathogenic." Su Dongpo writes accordingly, "He who knows how to nourish life will establish a good measure between work and leisure" (see Chu
1999, 71).

"Examining the proper state of practice, we should first of all consider the mind . . ., then focus on *qi*, look at the overall positions required,
and finally consider the specific aspects of the form." Movements and the
release of force have to be steady and proceed without disruptions, like a
silk thread smoothly unraveling, water softly flowing, or a wheel steadily
turning.

All movement, moreover, comes from one center, the elixir field, essential in Jiao Guorui's Practice of Spontaneous Unfolding as much as in
all other forms of qigong. The One in the body is the belly, and "guarding
the One" means cultivating Dao. In practice, this means, that all movements receive their impulse from the middle of the body. During practice,
we consistently breathe from the lower abdomen, the elixir field. Our in-
tegrity as oneness of body and mind/spirit accordingly arises through a
movement from the inside-out, making the body into a kind of three-
dimensional screen.

The notion of the elixir field originally goes back to operative alchemy: it denotes an area in the lower abdomen where *qi* is concentrated
and transformed. Qigong is a form of internal alchemy: we gather *qi* in our
elixir field and transform it from coarser into subtler forms of energy. This
is how we develop toward Dao.

Operative alchemists used cinnabar (HgS) as the prime ingredient of
their elixir of immortality. Through various chemical processes and by
mixing it with many different ingredients, they would transform or "re-
vert" cinnabar into the divine pill which would bestow immortality and
allow immediate ascension into the heavens .[24] Daoists were well aware
that the heavy metals contained in the concoction caused the end of the

---

[24] On "enhancing" and "eliminating," see Schubert (2000, 69). On the "proper
state of practice," see Jiao (1997b, 77). On "guarding the One," see Kohn (1989a).
On the elixir of immortality, see Darga (2004b, 90) and on transforming cinnabar,
see Needham (1976, 243-44).

physical body and led, in fact, to death. Internal alchemists and qigong practitioners do not face this problem: their elixir is internal and thus free from severe physical risks. The goal of qigong practitioners, moreover, is not immortality but rather entire harmony: they want to "bring all aspects of life into balance with nature . . . and attain oneness with Dao" (Bartl 2003, 69).

Integrative Interpretation

In qigong practice, it is "essential to create an integration of control and naturalness as well as to avoid understanding these two as opposites." I would like to expand on the application of Jiao's explanation on methods of stillness in terms of the "Chinese tradition of analogous thinking." The Taiji symbol (Fig. 1a) shows how things we tend to see as opposites in actuality form a whole. This also holds true for other areas of life, which brings me to the question of whether the opposites "quest for meaning" versus "quest for laws of nature" do not in fact form one whole? Can physics with its question of "How does this work?" and the quest for meaning with its fundamental issue of "Why is this so?" complement each other?

In trying to recognize the larger connections, to detect, like Faust, "the inmost force which binds the world, and guides its course," we have to look at the issues from different perspectives.[25] In contrast to Goethe's expression, which already integrates opposites, training in modern natural sciences teaches us to approach things exclusively from the external, the objective and reproducible experiment. Purely rational understanding cannot provide a meaningful answer; yet, being human, we continue to search for it. These days we have more rational knowledge than ever before and yet people seem to be more lonely and unhappy than ever. I personally do not believe that physicists will ever discover the meaning of life in a spectrometer—just as physicians could never discover the soul through dissection.

---

[25] On integrating control and naturalness, see Jiao (1997a, 15); on analogous thinking, see Jiao (1996c, 38). For Goethe's *Faust*, see www.gutenberg.org.

To reach a more profound understanding, then, we should approach issues not only from the scientific perspective, but also from the opposite direction, that is to say, from an internal, subjective, unique, and intuitive angle. Combining the results of both approaches into a new level of oneness should then, I think, provide a more complete level of understanding. What, then, is this internal, subjective approach? Human beings and the cosmos, as the *Yijing* already notes, are not at all separate. Is it, therefore, possible that a deeper level of awareness and understanding of self might reveal the nature of universal change? It seems to me that, among the various approaches that use a radically different direction than the natural sciences, mysticism is the most potent and informative. Modifying Jiao Guorui's statement cited above, I therefore propose the following thesis:

"To understand the connections in the universe it is essential to create a new level of integration of the natural sciences and mysticism while avoiding the perception of these two as opposites." If this is true, then mysticism and the natural sciences should form a single Dao, as much as yin and yang, stillness and movement are integrated in oneness.

Applying the known characteristics of the polarities, moreover, the following fundamental statements should hold true:

a) Each pole contains the seed of its opposite.

b) Both areas lend themselves to the building of further complementary pairs, i.e., pairs of opposites that form a whole.

c) Each area is in itself another Dao or Taiji.

d) The polarities can transform into one another.

Each pole contains the seed of its opposite (a): Research in the natural sciences also depends highly on intuition and inspiration, for example, when we interpret experimental results, such as Bohr's idea of complementarity in quantum physics, or when we plan new experiments. An example here is Descartes (1596-1650), whose "visionary enlightenment" at age 23 "determined the entire course of his life" (see Borchert 1997, 316). Despite this, he decided subsequently to study the world by strictly separating the internal from the external, which has led to the current worldview of the modern West.

Yet even in the area of mysticism we need an opposite pole, which here means formally-structured methods, clear and rational practice guidance—to prevent erring from the path when reason goes to sleep. The fundamental condition that each pole contains the seed of its opposite thus holds true for both areas.

Both areas lend themselves to the building of further complementary pairs (b): We find further complementary pairs as soon as we assume that the human quest for meaning and understanding has a long history. How did people deal with this in the old days? My personal vision has human beings in hunter-gatherer societies several thousands of years ago facing natural events with astonishment and awe (see Fig. 4). They were curious and courageous, so that the initial astonishment soon led to analytical, proto-scientific approaches while the attitude of awe led to religious concepts and mystical experiences.

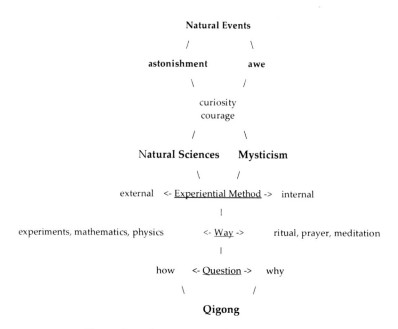

Fig. 4: Complementary methods to acquire information

The natural sciences and mysticism are both experiential methods, using external and internal approaches. Their ways of acquiring information include experiments, mathematics, and physical theories, on the one hand, prayer, meditation and ritual, on the other. Obvious questions refer to function (How is it that the sun rises every day?) and meaning (Why am I here?). As shown in Fig. 4, there are numerous pairs of opposites that build one whole. Both areas thus encourage the building of further complementary sets.

All this also applies to qigong, where I see a clear connection between the two areas. Qigong involves a method part (conditions) and an experiential part (practice). Each successful qigong exercise combines opposites in one whole. For example, in the third of the Eight Brocades, "Uphold Heaven and Push on Earth," we stand in parallel stance and, after an initial rising and descending of the hands, let them separate. The left hand rises overhead to uphold heaven, while the right hand descends to the right hip to push down on earth. This is the opening part of the movement. To close, we let both hands approach at shoulder level, then open them into a big circle and move them down to end in front of the lower abdomen. Here the opening part reflects the two polarities, which are further pulled together in twofold fashion during the closing section.

Qigong thus provides a concrete, tangible, bodily experience of how the oneness of the opposites feels. This brings us, I think, already a step beyond Faust's dialogue partner Wagner. He tells him, "Unless you feel, naught will you ever gain."[26] Just as Zhuangzi appreciates the "happiness of fish" through his own experience of happiness while walking near the river (ch. 17), so we also get a glimpse of the oneness of Dao in every little practice exercise. The fundamental principle of practice, as Jiao notes, is "the ongoing mutual transformation of yin and yang" (1993a, #61). Besides its immediate, tangible effect in practice, there is also the scientific foundation of qigong: it has a long tradition of detailed descriptions of practices, and there are increasing numbers of studies that show its potent effectiveness for treating a variety of conditions. As we practice qigong, while paying close attention to wholeness, we find the integration of opposites into oneness in each and every part of the practice. We thus get to experience a small portion of universal oneness, a quantum of Dao.[27] Can we, then, ex-

---

[26] For a description of the Eight Brocades, see Bock-Möbius (1993, 85-101); Jiao (1996a). For Goethe's *Faust*, see www.gutenberg.org. For the tradition of detailed description of practices, see (Kohn 2008).

[27] On the scientific foundation of qigong, see Jiao (1993a, #9; 1992, 262). On the "quantum of *Dao*," see Bock-Möbius (2010a).

pand on this experience? Can qigong help us to overcome the polarity between the natural sciences and mysticism?

Each area is in itself another Dao or Taiji (c): Does this mean that even the natural sciences in themselves or mysticism in itself become yet another image or replica of Dao? I will come back to this complex point later in the discussion.

The polarities can transform into one another (d): My first thought on this fundamental condition is a rather humorous one. Even highly rational scientists may on occasion become irrational when they do not know how to proceed. Such scientists may say, for example, "Homeopathy is quackery." When asked whether they have ever tried it and what its fundamental concepts are, the irrational aspect surfaces, "Of course not! We don't touch such nonsense!"

On the other hand, there are people who are considered sages. They live and think simply, presenting clearly formulated and easily acceptable principles of living. An early example appears in the *Zhuangzi* in the figure of Cook Ding, who is so deeply in synchronicity with the structures of an ox's carcass that his knife remains forever sharp (ch. 3). Zhuangzi presents several examples of such ordinary people, who live simply and adapt their natural way of being to the circumstances of life as models. They are characterized as having deep understanding because they are aware of their own simplicity and are neither proud nor presumptuous. It is, therefore, quite possible "to be wise without extensive knowledge" (ch. 4)—simple and without intentional action, like the hole in the wall that opens the entire room to the light. Nonaction thus means living just so. As Bai Juyi 白居易 (772-846) notes in his poem on "Sitting in Stillness": "I sit! Trusting the sun, my eyes closed!"(Dahmer 2007, 106).

All these considerations document the validity of my fundamental approach, which is to no longer see the natural sciences and mysticism as opposites, but instead as complementary areas, as varying facets of an integrated wholeness. The next chapters present more details on just how wholeness and oneness appear in both mysticism and quantum physics.

Now take some time to digest the theory, while enjoying the first set of exercises (see ch. 7 below): the preparation and the first three exercises from the First Cycle, "Origin."

4. Mystical Union

How can I tell you about the colors of the sky,
when you don't know that there is a sky?
Or about the wind that carries my arms,
when you don't believe in the wind?

How hard is it to speak of oneness even after one has experienced it? What meaning, what relevance, can we attribute to internal experiences? There are numerous reports of people who, under a certain level of tension or during difficult situations sent off a prayer, a plea, a petition and had the feeling that they were heard. They claim that they received some kind of an answer, and therefore felt they were part of something else, something bigger, something that carried them along. We may all have felt something along these lines on a smaller scale, but mystics explain how it feels on the larger, cosmic level.

What Is Mysticism?

Qigong practitioners call the overcoming of polarities, the overarching sense of great wholeness - the experience of Dao. Mystics call this "union" and describe it in quite similar terms. Thus, Meister Eckhart notes, "When the soul comes into the light of reason, it no longer knows of opposites" (Quint 2007, 194).

What then do we mean by the term "mysticism"? It is certainly not just a general term for all kinds of nebulous ideas and visions. According to Willigis Jäger—Benedictine monk, Zen master, and mystic—mysticism is "a term for trans-confessional spirituality" (2000, 59), an understanding gained through introspection rather than by learning and thinking. The term, moreover, goes back to ancient Greek and originally indicates the closing of eyes and lips. This feature also appears in the *Zhuangzi*, where Laozi describes the unitive experience thus, "My soul is tied and cannot think; my mouth is closed and cannot speak" (ch. 21).

Mysticism thus indicates a form of spirituality beyond religious orthodoxies, beliefs, and confessions. It connects directly to awareness and leads to the experience of the absolute in the here and now. It means reaching an "accord with the underlying law of the cosmos" (Jäger 2005, 155), an "experience of the highest mystery" (Swami Sudip 2004). This understanding of spirituality sees the mind as the highest reality and considers matter as one of its manifestations, as mind made dense.

As shown in Fig. 5, Jäger assumes that religions share the same goal on the level of inner experience, i.e., in their esoteric dimension: they come ever closer together until they eventually join in a single point. Despite this ultimate unity, each religion defines the path to oneness with the highest power in its own unique way. Each expresses the experience in words and images, its more exoteric dimension, on the basis of its particular cultural context, resulting in numerous different creeds and orthodoxies (Jäger 2005, 39).

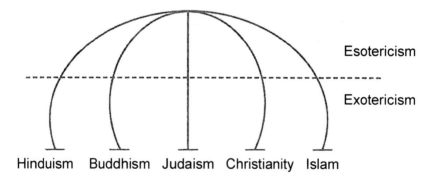

Fig. 5: Differences in the exoteric dimension disappear at the esoteric level

The experience itself is clear and unmistakable, but its telling depends on words and concepts, making it a difficult undertaking indeed.

Personally, I see the approach to the highest power as depicted in Fig 6. The various religions show us the general direction to the absolute, but they cannot overcome a certain border. The highest point cannot be reached with rational means: undefinable as a matter of principle, it has to be experienced to be known. This situation is not unlike that of modern cosmology: the question about what was before the big bang has no meaning, because neither space nor time was defined then. Still, we want to know how it all came about. Something arose that I will call the principle of "life;" we called it Dao or *qi* earlier.

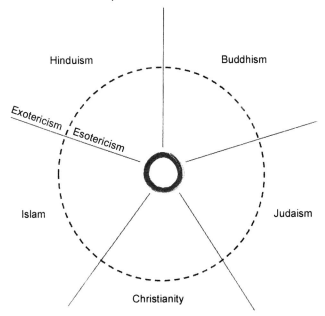

Fig. 6: Approaching the absolute

Willigis Jäger, in his books *Die Welle ist das Meer* and *Suche nach dem Sinn des Lebens* (*Search for the Meaning of Life*), presents beautiful descriptions and pictoral comparisons to outline what cannot truly be described. Even though his efforts remain circumstantial, representing shadows of a deeper truth, they make us curious about the path (1991, 15). Diving into transpersonal oneness is what we call "mystical union": if God is the ocean and human beings are the waves, then mystical experience means that the wave realizes its true nature as part of the ocean—while simultaneously experiencing the ocean as wave. Both are at the same time ocean and wave.

Generally religions can be seen as consisting of theology (their rational part) and mysticism (their experiential dimension). In Catholicism specifically, mysticism was classified as part of official dogma, which meant that all reports about experiences had to be phrased according to the rational teachings of the faith. As a result, they were strongly filtered and came to play only a secondary role, even to the point that the theologians forgot all about mystical experiences. Willigis Jäger in his works strives to recollect and revive the various mystical traditions, especially also those that form part of Christianity. He is not alone in this effort. As the renewal theologian Karl Rahner notes, "The Christ of the future will be more mystical." Jäger expands on this, "The human being of the future will be more mystical" or s/he will not be at all (2000, 30). The philosopher and theologian Titus Brandsma, President of Nijmegen University, declared as early as 1932 that "mysticism is necessary to solve the world crisis."

The experience and the potentiality of experience of divinity beyond pure faith is the position from which the question of meaning can ultimately be addressed. God, often also called prime reality, "is seen as the one reality that reveals itself in many forms yet always remains itself. This reality is like the ocean that reveals itself in millions of waves yet always remains the same water."

Experiencing unity means overcoming the ego while uncovering the Self. The ego is the organizational center of our personality structure, which manages our daily life in the world. The mystical experience teaches us not to identify with the ego and instead to recognize connections. The mystical world of Hildegard von Bingen (1098-1179) accordingly was based "entirely on the personal experience of a meeting with God" (2001, 19). The more we get our ego to step out of the way of God, the better He can work through us. This is very much like nonaction: as we do not obstruct or interfere with nature, there is nothing that is not done (Daode jing; 37). Thus, the sage joins Dao.

All religions are paths to experiencing the divine—there are many of them, but they all lead up the same mountain. Jäger regards them like church windows, "They provide a certain structure to the light that shines through them" (2000, 47); without light they are dull. Just as church windows give color to the light, so religions give expression to the ineffable. To put it another way: God appears in the various religions like light in colors of the spectrum. This image makes it clear that each individual or group may have their particular favorite color, yet no color is better than another.

Nevertheless, religions also have their shadow sides. In that respect, they are like the moon, which illuminates the earth at night, but takes its light from the sun. As soon as the moon moves between the sun and the earth, there is an eclipse and there is darkness. In this metaphor, the sun represents the divine, while the earth stands for humanity. Should religion (the moon) move between God (the sun) and humanity (the earth), it would cover God, producing an "eclipse of divinity" for humankind. In other words, religion would occupy an inappropriate place that would hinder rather than open up the experience of the divine. Yet despite these potential issues, religions provide a major resource for humanity to reach toward spirituality.[1]

The Separation

In Christianity there is a gap between God and world. As the world's creator, God is fundamentally separate from us, which means there is no original oneness, and not even unity in the times of first creation. After that, the connection to God is established through formal covenants (with Abraham or Moses) and a savior (Jesus Christ). Christ shows us the path toward the experience of oneness. He knows the way because He has walked it Himself. Our sense of reason tends to like the concept of separation: familiar and well accepted, it matches the inherent, dualistic structure of the rational mind, which essentially recognizes through making distinctions. Eastern traditions take a different approach to this: they propose an original state of oneness which is never lost; the divine accordingly opens

---

[1] On the wave-ocean parabola and the "eclipse of divinity," see Jäger (2000, 7, 42); (2005, 39). On mysticism to solve the world crisis, see Borchert (1997, 315). On the ego and personality structure, see Jäger (2000, 179), Borchert (1997, 319).

like a fan in the evolution of the world - all its folds forever connected and part of the One.[2]

Western mysticism has continued to wane, especially also due to the increased impact of science on modern theology over the past two centuries. Spiritual exercises have been reduced to "discursive meditation" and there is no longer instruction in the practice of contemplation. Generally, the mystical approach is instead rejected, and many theologians have a predominantly intellectualized image of God. Meister Eckhart already warned against a God that one only thinks of and does not feel, "As soon as the thought is gone, so is God." The idea of God beyond and outside of creation is also part of Greek thought, notably of Aristotle. His teacher Plato, on the other hand, did not follow the dualistic-theistic approach: he assumed that each person could find the fundamental ideas of the universe in his or her own soul—a soul that served as the mediator between the world of ideas and that of meaning. For him, matter only becomes reality due to the ideas that are themselves beyond time and space.

Many people living in the rationally-oriented world of today never even notice their spiritual needs, or if they do so, do not take them seriously. To me, this is one of the main reasons for the crisis of meaning in modern society, an expression of the strong awareness of existential separateness. The Austrian singer Claudia Mitscha-Eibl describes this, "It's got to be silent in heaven, no word penetrates the silence. Who used to live there—is he dead? Or did he just walk off?" (see Jäger 2000, 50)

Still, Meister Eckhart assures us, God will not remove Himself any further than right in front of the door if He is not allowed to remain inside. The poem "Traces in the Sand," too, enhances this understanding: Jesus is not someplace far off in Heaven, but right with us. In those times, when we feel left alone and only see one set of tracks in the sand instead of two, those are the tracks of Jesus as He carries us along (see Powers 1996).

Modern life runs differently: most often we focus on what we represent, thus strengthening characteristics that support a strong ego, which is the exact opposite of "Thy will be done," and that despite the fact that we

---

[2] Regarding Christ's path, spiritual exercises and the "divine fan," see Jäger (2000, 20, 118; 2005, 42).

all have the quest for transcendence within us, "The soul is . . . in misery until it . . . reaches the eternal good . . . it was made for" (Meister Eckhart). The common tendency today is that we never take the time to even perceive ourselves as seekers. Yet even if we never permit this quest to play a role in our lives, it will still push its way in. We may then perceive it as a restlessness that requires immediate relief. We give it a quick fix, which cannot fulfill our needs. This, in turn, may lead to addictive behaviors— the visible manifestation of a quest gone wrong or cut off too early (Dethlefsen 1983, 330-31). Ego-fixations, self-attachments, and avoidance of transcendence and spirituality are ways of rejecting integration into a greater reality. They indicate a turning-away from the core meaning of life. Carl G. Jung already realized just how much people suffer from their failed struggle for meaning, "Among all my patients beyond mid-life, i.e., after 35, there isn't one who does not face the ultimate problem of his or her religious attitude."

The biblical story of Genesis tells of humankind's exile from Paradise after Adam and Eve tasted the fruit of the Tree of Knowledge, thus "falling into sin." This is its way of describing the loss of original oneness. Neutrally evaluated, this means: falling into sin led to the awareness of the separation of God and the world, as well as of good and evil in all their variants. Good and evil only coincide again in prime reality (Borchert 1997, 278).

The notion of sin in this context means that people have come to see themselves as separate from God through becoming ego-focused. The key task, therefore, is to transform our perspective. Our deepest yearning is God, and God wants more than anything to unfold within us. This yearning, then, shows us the way through the pain and suffering of evolutionary unfolding. "Original sin is not guilt in the narrow sense of the term; . . . rather, it is a fact arising from the development of human consciousness." "We never completely fell out of God." Instead, in our state of separateness, we have merely forgotten where we came from and are not aware that there would be no development of life without our entry into polarity.[3]

---

[3] On Meister Eckhart, see Quint (2007, 60, 78, 380-81). On our personal quest, see Jäger (2005, 131, 80, 45).

Ken Wilber calls original oneness "prepersonal heaven." This refers to that state of dumb preconsciousness before humanity perceived its existence as separate from nature, the state from which we have gradually arisen as shapers and transformers of nature. For most of human evolution throughout history and until today, we have continued to rise above, and become independent from, nature. However, only by overcoming our current reckless and egotistic way of living in favor of a transpersonal consciousness can we recover heaven (and the world) in a healthy manner — by reaching a conscious level of harmony and a state of oneness with nature.

This process, however, can never just go back to the prepersonal state: that would be mere regression to stages of consciousness that we have already outgrown. Ken Wilber, the leading theorist of transpersonal psychology, discusses this in *Up from Eden* (2004) and *The Atman Project* (1990).

He begins by presupposing that history has meaning, seeing it as a "report on the relationship of humanity to its deepest being" (2004, 26). In other words, he sees all development as goal oriented and leading toward oneness with the highest creative principle. Overcoming the separation and "rediscovering wholeness" are the key requirements and core desires of humanity (1990, 170). This desire can be fulfilled through mysticism, without postponing it into the otherworld (Jäger 2005, 114). According to Wilber, there are only two states that see humanity happy: either the "sleep of the unconscious (prepersonal level) or "awakening of the superconscious" (transpersonal level).

At the present time, we are halfway in-between with self or ego consciousness (personal level; 2004, 136). This is made easily visible in Fig. 7, which shows the different sectors in the Great Chain of Being. Humanity's psychological development is the microcosmic mirror of macrocosmic unfolding of both humanity and the universe, best depicted in stages 1 through 8. It has the same goal as natural evolution: to continuously unfold into higher levels (1990, 167). Hildegard von Bingen says it slightly differently, "The entire world . . . joins humanity on its way to salvation" (2001, 184).

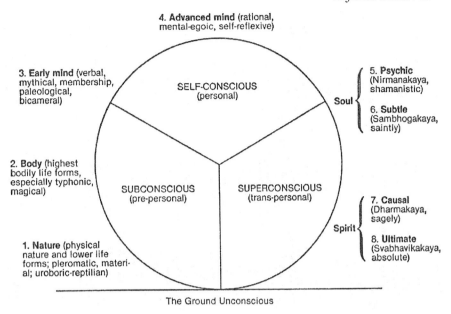

4. **Advanced mind** (rational, mental-egoic, self-reflexive)

3. **Early mind** (verbal, mythical, membership, paleological, bicameral)

SELF-CONSCIOUS (personal)

Soul {

5. **Psychic** (Nirmanakaya, shamanistic)

6. **Subtle** (Sambhogakaya, saintly)

2. **Body** (highest bodily life forms, especially typhonic, magical)

SUBCONSCIOUS (pre-personal)

SUPERCONSCIOUS (trans-personal)

Spirit {

7. **Causal** (Dharmakaya, sagely)

8. **Ultimate** (Svabhavikakaya, absolute)

1. **Nature** (physical nature and lower life forms; pleromatic, material; uroboric-reptilian)

The Ground Unconscious

Fig 7: The Great Chain of Being

Wilber describes this development in two parts. The first leads from unconsciouness to ego-consciousness (Stages 1-3). It includes the early stages of life: newborns, who do not experience themselves as separate from the mother (left section), then gradually grow, leave the oneness with the mother, and emerge as their own persons (Stage 4, upper section). The second part leads from ego-consciousness to superconsciousness (Stages 5-8, right section). It indicates the advanced stages: people overcome their ego-identity, reach transcendence, and find their true Self in transpersonal oneness. Jäger formulates it like this: At this point, humanity has opened up the transpersonal as an additional dimension of reality (2005, 104).

Since new abilities do not come from themselves, they must, Wilber points out, have been present before, but were hidden or inactive. In explaining this, he does not see the beginning of evolution about fourteen billion years ago as the original beginning of everything, but as the end of an "involution" that occurred before it (see Fig. 8). Involution means the forgetting or folding in of higher levels. This vision implies the common position that reality is hierarchically structured, from the more material to the more spiritual.

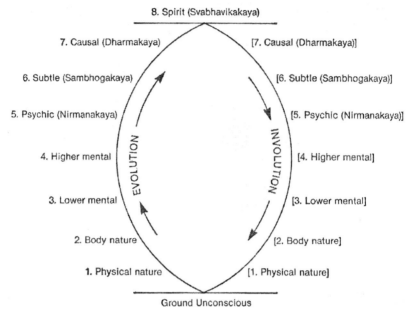

Fig. 8: Involution and Evolution

The latter is called Brahman in Hinduism—it stands in a twofold relation-ship with the universe: involution, the folding in of higher structures—Brahman steps outside of itself to create the world; and evolution, the un-folding of hidden levels—the world again moves toward Brahman. At the end of involution, all higher levels are unconscious; the sum total of the unconscious is what Wilber called the "ground unconscious" (1990, 263; 2004, 344-45). It is from here that evolution begins, a gradual re-collecting of nature into God, a re-unfolding of all that is potentially present to the eighth stage of "spirit."

As qigong practitioners, we do something quite similar, albeit solely for ourselves and not on a cosmic plane: we mobilize latently present forces. These are forces that are already present deep within us, but which only unfold through practice.

This model of a circular process between involution and evolution, moreover, contains the possibility of reconciliation between the natural sciences and religion, inasmuch as the well-known evolutionary process can be integrated into a combined picture. In addition, it also contains an answer to the still unexplained question of how exactly new faculties develop.

Our task in this life is to recover our original core. Unfortunately, however, we do not have "direct access to the core of being" (Jäger 2005, 90), and the way there by necessity passes through the overcoming of the ego. Since this process is painful, we try to evade it, which in turn leads to more suffering and further detours. Whatever we consider our "ego" is just a mask, a role we play in society. Our true core lies behind it. For this reason, our limited self, our ego, has to die, and only then can we find mystical union, become one in our souls with God.

We yearn for transcendence, yet we are afraid of the death of the ego and thus try to find replacements. This, in turn, leads to a variety of faulty developments: desiring to reach timeless wholeness, we strive for immortality; searching for oneness with the cosmos, we start to run the universe; hoping to become one with God, we begin to play God ourselves; driven by a quest for the infinite, we mistake it for the finite. Instead of becoming one with God, we look for God in material things and make money into a symbol of immortality. Instead of pursuing harmony with nature, we push for the conquest of outer space. All this entails a fundamental separation of nature and humanity, of body and self.[4] This has been most pronounced since Descartes in the seventeenth century, when science and mysticism separated, when research was objectified and thus no longer directly connected to the explorer, when alchemy became chemistry and science only focused on external reality.

Overcoming the Separation

Mystics see everything that is as a "manifestation of God" or, as Meister Eckhart says, "all things . . . taste of God." A very similar sentiment also appears in a Tang poem by Bai Juyi, "Pondering the Taste of Dao." In mysticism, there is no separation of God and the

---

[4] On replacements and faulty developments, see Wilber (2004, 28, 220, 325). On alchemy and chemistry, see Borchert (1997, 318).

world. The separation of God and the world is just as impossible as the division of yin and yang or the attempt to remove information from a hologram.[5]

Following Jäger and Wilber, the future of religion is in integral spirituality, already described in the perennial philosophy as expounded by numerous important spiritual teachers. The perennial philosophy presents the basic tenets of mystical spirituality in intellectual and rational concepts. In content, it can be traced back as far as Neo-Platonism of the third century, as founded by Plotin and Proclos. The term itself was coined by Eugubinus around 1540 and later adopted by Leibniz (Jäger 2005, 106).

Its fundamental tenets can be found in practically all cultures and in all periods of history, always with the same essential teachings. They form the core of the various mystical teachings of the world and decribe the underlying oneness of religions. Its key concepts include: God exists and can only be found within; living in a world of separation, we fail to realize the mind within. However, we can liberate ourselves from the state of separation through meditation, concentration on the breath or other focal point, and thus experience this mind directly. This, in turn, means the end of suffering, leading to compassionate action in a timeless now.

Overcoming the separation can occur through mental work, such as contemplation, as well as through the body. Unfortunately, Western culture does not have a religious tradition that is fully at home in the body, despite the fact that the body is much closer to our true being than reason. Most other cultures acknowledge this fact, but Westerners deny it for the most part, using only their intellect to reach transcendence. This is like trying to play the piano and only using one octave, like facing a poem without knowing how to read: we count the words and letters, but have no clue about the meaning. The theologian and psychoanalyst Eugen Drewermann provides a particularly strong image, "In everyday consciousness, we are like someone sitting in a fish tank. Convinced that the

---

[5] On mystics and mysticism, see Jäger (2000, 48, 19); on Meister Eckhart, see Quint (2007, 60). For Bai Juyi's poem, see Dahmer (2007, 102).

glass partition signals the end of the universe, we are frustrated and depressed because the limitation fails to offer meaning" (Jäger 1991, 235).[6]

Praying means talking to God; but even older than praying are prayers without words, prayers of the body. Becoming one with gesture and movement leads to a deep level of concentration and opens the transition to the transpersonal level of consciousness. "Religious gestures awaken religious feelings within us. Reminding us of forgotten levels of being, they create an openness to the transpersonal dimension and may well become . . . a prayer without words." Being in the right gesture, moreover, means that one is in the world in the right way. We are nothing but an expression of God, divinity that has taken shape. Physical gestures are thus not only an expression of the connection with the divine, they are the very connection itself. Christians have for too long defamed the body as an "obstacle to the mind," regardless of how much even great mystics like St. Teresa of Avila (1515-1582) have admonished, "Be kind to your body, so your soul enjoys living in it."

Besides body prayer, there are several other forms of meditative movement that may lead to higher levels of consciousness, depending on what intention we place on them. One example of this is found in pilgrimages: in walking from holy place to holy place, we meet God in our bodies; we walk for the sake of walking, not primarily to arrive anywhere. Along the same lines, God walks in us and through us on this earth, while our body is "like an instrument that resonates with God." He reveals himself in us as human being: we are God, but only in the experience of oneness, not in our separate ego consciousness. "Knowing God does not lead to knowledge of God but to oneness with God."[7] This is also obvious in the deep visions of Hildegard von Bingen, who called herself "an uneducated woman" (2001, 26).

The main task we face as humans is to become fully human or, as qigong masters express it, to learn "the principles of truly being human" (Jiao 1993a, #13). This means overcoming the ego and stepping into trans-

---

[6] On the perennial philosophy, see Wilber (2004, 16); (1996, 99-100); Huxley (1946); Forman (1990). On the limits of the intellect to reach transcendence, see Jäger (2000, 31, 85; 1991, 235).

[7] On body prayers, see Jäger (2000, 128, 131; 2004, 24, 30, 44, 70; 2005, 118).

personal consciousness, thus fully overcoming the separation of God and world. This will finally open up true reality to us, allowing us to find our true "identity with prime reality" (Jäger 2000, 33-34). This level is, moreover, far beyond any levels of the subconscious attained through psychoanalysis or psychotherapy: it contains a huge potentiality for healing. Being healed, then, means to have understood the true meaning of life—which, in turn, means that even a person who is physically sick may well be healed.

Experiencing the transpersonal level causes a transformation of personality that is inaccessible to resolutions or will power. The individual perceives deeper connections of human life and enters a state of consciousness in which he or she is no longer a mere ego, but purely awake and present. Awakening, clarity, and presence are similarly the goals of qigong practice and are described in similar terms. The ego is secondary; it is not about how good we look during practice, but to what degree we embody and realize the core principles that are founded on the ongoing transformation and ultimate oneness of yin and yang (Jäger 2000, 155; 1005, 52).

As Willigis Jäger notes, people undergoing a mystical experience also realize that there is no death; instead, there are only the beginning and end of forms or expressions, leaving prime reality untouched. Any fear of dying fades away since, as a mystic, dying means "dying into something much bigger." Why, then, should we be afraid of the "ship sinking as long as God is the ocean into which it sinks" (2000, 178)?

Most people find it hard to integrate an experience of oneness into their lives; they feel thoroughly rattled and deeply insecure (2005, 110). They need support in interpreting their experience inasmuch as the confusion that often comes from a spiritual experience can also lead to a profound fear of the ego of losing control in the transpersonal space. Starting from the personal state, both psychotics and mystics have the same goal of becoming one with the highest principle, but they pursue it in opposite ways. Mystics move forward on the transpersonal way of greater unfolding; psychotics try to move back into the mother's womb, pursuing a strategy of regression that is unwholesome. That is to say, mystics, unlike psychotics, do not sink in the ocean of consciousness (2000, 163).

Ken Wilber sees mystics as embodying the highest level of being human. How, then, can we distinguish the prepersonal from the transpersonal yearning? How often are the spiritually advanced mistaken for the mentally disturbed, and vice versa? Is it possible to give guidance toward a more positive unfolding to a psychotic person? Would that open his way toward oneness and health? Experiences of oneness are not as exceptional as one may think. Many philosophers in the past had them— Schopenhauer, Nietzsche, Descartes, Heidegger, and Jung, for example.[8]

I have spoken much about God in this chapter. It is, however, entirely possible to replace the personal God of Christianity with a supra-personal principle such as Dao, to think of the struggle between good and evil in terms of the ongoing transformations of yin and yang. Doing so, I think, leaves no contradictions between these reflections on mysticism and the explications on qigong.

Traditionally, of course, this was not acceptable: in 1329, Pope John XXII condemned Meister Eckhart's explications that God was beyond all characteristics, and thus not essentially good, as heresy (Quint 2007, 454).

As qigong practitioners, we have to pay particular attention to maintain complete control over the practice situation, especially also as we approach the experience of oneness. It is important to remain lucid and maintain the state of awakening, clarity and presence even as we realize that we are merely part of the greater ocean. I think this is the critical point. The consequences of what happens when presence is lacking is most obviously made visible in Goya's painting "The Sleep of Reason" (see Fig. 9).

"The mystical process has its own inner compass." It is essential to maintain one's direction, while moving through the labyrinth of life (Borchert 1997, 363). That is to say, we continue to have a specific place as individuals in the world, even while remaining aware that we, as well as the entire earth, are nothing but a speck of dust in the greater universe.

---

[8] On spiritual experience, see Wilber (1990, 132); Borchert (1997, 24); Jäger (2005, 110-12).

Fig. 9: Francisco Goya – El sueño de la razón produce monstruos

(The sleep of reason produces monsters)

Applying proper care and caution there is no reason why we should avoid the inner experience and limit ourselves to reason on our quest for deeper understanding. As has become clear here, mysticism is also a Dao in itself: next to inner experience, we need to keep up our clear presence. This also means that the earlier thesis regarding the attainment of inclusive understanding through combining the natural sciences and mysticism is fulfilled with respect to the point that each area is a Dao in itself, "Divine nature is one; each person is also one—the same one as is nature" (Quint 2007, 145). The transformation of the polarities, moreover, is evident in the mutual conditioning of inner experience and clear presence (Bock-Möbius 2010b).

Before we move on into physics to appreciate the principles found valid for qigong and mysticism in their nature science guise, I recommend a break for practical application: exercise 4, "Reverting to Original Light," in the First Cycle, "Origin" (see ch. 7 below) opens an experience supporting what has been discussed so far.

5. Entangled Reality
Nonlocal Interactions in Quantum Physics

That we in truth can nothing know!
That in my heart like fire doth burn.
—*Faust*

I stumbled over several serious, deep questions in my fourth year of graduate studies in physics. It was the last lecture of my last summer term. While teaching us about nuclear and elementary particles, a young, enthusiastic professor explained the underlying principle that gives order to the Baryon Octet, i.e., an assortment of elementary particles that consists of up, down, or strange quarks as their building blocks. This ordering principle, collectively called the symmetry group SU(3), had been presented in 1961 by the inventors Gell-Mann and Ne'eman under the name "The Eightfold Way." Alluding to the Noble Eightfold Path of Buddha, which liberates those who walk on it from suffering, the Eightfold Way in elementary physics alleviates the physicists' suffering from innumerable minute particles, which appear not to fit into any existing ordering scheme. With this statement, the professor concluded his course and wished us a happy summer vacation . . .

I didn't quite know what I needed to ask, but I felt strongly that things had been left unfinished: this could not possibly be the last lecture of the term. It felt as if this lecture had been the very beginning, the first introduction to a topic that was of burning interest to me. However, I had already completed my main program of study at this point and had been exposed to a little bit of everything. I was now ready to start writing the thesis for my diploma.

Despite this, I felt like I had just listened to my very first lecture: I had no clue about the topic he had just addressed. So the question arose: what to do about this Eightfold Way? I asked the teacher. He suggested I could read about the teachings of Buddha; everything else was just metaphysics. That was all the friendly professor told me. And what did he mean? Perhaps he was thinking of the definition of metaphysics in Goethe's *Faust*: "See that you most profoundly gain what does not suit the human brain." Was that it?

I could not possibly put the topic to rest. Something deep inside me was resonating with it and longed for expression. Up to that point, when I had come up against boundaries, the issue had always been my own ignorance. Now suddenly I felt as if there were certain questions in physics and the natural sciences that people intentionally refrained from asking, as if there were things that nobody wanted to acknowledge. But no, this couldn't be it, could it? I wanted to "detect the inmost force which binds the world, and guides its course" (*Faust*), and in this quest should refrain from asking inward questions?![1]

As I pondered these issues, the vague idea began to take shape that I might spend some time after my diploma in India or China to gain deeper, and possibly completely different, access to this idea of the "way/path." This became reality in 1984. I flew to China as a tourist, but wasn't yet satisfied by the impressions I gathered. As a result, I decided on another visit, not as a tourist, but to live and work there. This plan materialized in 1988. Following my stay at that time, I continued to study and explore the Path for over twenty years. Now the main task is to bring both dimensions of my inquiry—physics and qigong/spirituality—into harmonious integration.

To this end, I will now turn to various physics experiments and theories I find best suited for this purpose. The topics may seem complicated, but there is no reason to be scared of the unknown. Just follow the main lines of argument, which is somewhat of a pedestrian approach to quan-

---

[1] For an edition of Goethe's *Faust*, see www.gutenberg.org.

tum physics that should be as much fun as work. I will include some technical formulas to summarize the discussion in the text, but these are separately marked and can be skipped at the reader's discretion.

Why Quanta?

S o far, we have focused mainly on physical and spiritual practices, health enhancing and transpersonal techniques, as well as the experience of oneness, the process of awakening, and the ideal of being present. Meister Eckhart advises that we should not strive for anything but God.

This may sound great, but is not so easy, since everyday life presents us with all kinds of tasks, and we inevitably live in the material world—a world dominated by rational thinking, the need to objectify, and the demand for reproducible results. We measure, weigh, observe, and analyze to understand connections. Everything works out nicely as long as we concern ourselves with objects and systems on scales we are used to in everyday life. However, this changes radically once we dive into the atomic scales. The Newtonian worldview, which describes the world as a mechanical clockwork and which has been around since the seventeenth century, was shaken to its foundations about 100 years ago. Why? And what does this have to do with qigong practice? To answer these questions, let me wander a bit farther afield to eventually return to the topic of oneness from the perspective of modern physics.

Modern quantum physics began on October 14, 1900 at a congress of the German Physical Society in Berlin where Max Planck (1858-1947) presented his new theory. It was born out of desperation in Planck's attempt to derive the relation that describes the radiation energy of a "black body": a cavity with a small opening that absorbs all incoming radiation and transforms it into heat. The emitted radiation exclusively depends on the body's temperature. However, the experimental results concerning the emitted radiation did not match the hitherto known equations.

Faced with this dilemma, it occurred to Planck that perhaps not only matter comes in smallest units (atoms), but that there might be also something like an "atom of radiation," in other words, a smallest quantum of radiation that can be emitted. Originally, he introduced this quantum hypothesis only as a mathematical trick to simplify calculation, or a help factor (h), that he hoped to eliminate in the end. He thus made the assumption that radiation energy could only be released in integer multiples of this fundamental energy quantum.

This is expressed in the equation $E = h\nu$, in which "h" is Planck's constant, also called the "quantum of action" ($h = 6.6 \cdot 10^{-34}$ Js), and "ν" the frequency of the radiation. An action of this order of magnitude is extremely small, and it is obvious that its effects are vanishingly small on scales of everyday life. Thus, quantum mechanics was born.[2]

Let me illustrate the importance of this concept with a familiar example. Imagine a large pot of soup instead of the black body and let us examine the method of serving the soup instead of the emission of energy. We could pick up the pot by its handles and pour the soup directly from the pot into soup bowls. This would be a continuous serving of soup—comparable to a continuous emission of energy.

Alternatively, we could also take a ladle and fill the bowls in this way. In this case, the soup would be portioned out, i.e., served ladle by ladle. We can call this a discrete serving of soup. Now let us go back to physics and replace the soup with energy and examine discrete emission of energy. In other words, Planck's key assumption was that radiation energy could only be emitted in discrete amounts, i.e., in fixed portions, or quantum by quantum.

The most important thing about this hypothesis is not only that energy can be emitted in quanta, but also that there is a smallest portion that can be transferred, a minimum-size ladle, so to speak, for serving soup. For radiation, the size of this ladle is "h." Using this approach finally facilitated a description of the black body radiation, or energy emission, over the entire frequency range with a single formula. Only much later, under consideration of the nature of light, did it become obvious that the quantum hypothesis would have much farther ranging consequences.

---

[2] On Meister Eckhart, see Quint (2007, 83). For the Newtonian world view, see Schäfer (2004, 13). When and how Planck developed his new theory, see Audretsch (2002, XIII); Tegmark (2001, 70); Planck (2008).

Wave-Particle Dualism

O ver the past few centuries, various theories have been proposed about the nature of light. Isaac Newton (1643-1727) spoke about it in terms of corpuscles; Thomas Young and his double-slit experiment in 1803 established its wave nature, echoing the phenomenon of interference, already well-known from water waves.

In 1905, Albert Einstein (1879-1955) used the quantum hypothesis to explain the photoelectric effect. This arises when a negatively charged metal surface loses its charge while being illuminated by UV-light under vacuum conditions (see Fig. 10).

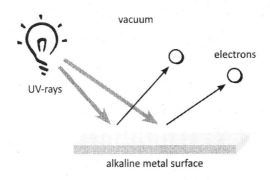

Fig. 10: Schematic view of the photoelectric effect

The high-energy light rays in this experiment quite obviously again behave like particles, expelling the electrons from the metal: these light particles are called photons. The photon energy depends on the frequency of the incoming UV-light. If the frequency is too low, however, no electrons are released, since the energy of the photons ($E = h \nu$) cannot overcome the electron's binding energy. With the explanation of the photoelectric effect, Einstein confirmed the applicability of Planck's constant.

The principle of the double-slit experiment is illustrated in Fig. 11. Wavefronts coming from the left hit a wall with two holes: the elementary waves that emerge from there, described by the wave functions ψ_1 and ψ_2, overlap and result in the intensity distribution P(x) on the screen as depicted. The experiment can be performed using either electrons or light: both produce interference patterns, thus showing that both possess wave character. Obviously, we associate the particle character with electrons, and it becomes of utmost interest whether one can, as in the particle picture, distinguish through which slit they passed before they hit the screen.

Consequently, experimenters wished to understand this behavior better and to observe the electrons on their journey through the slit using, for example, a light source between slit and screen. They found that in this case all interferences vanish and electrons appear as particles. This was extremely strange and caused many physicists serious headaches. Einstein noted that they were like ghosts: one fears them in the dark, but they vanish as soon as the light is turned on.

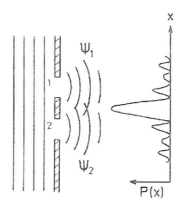

Fig. 11: Interference pattern in the double-slit experiment

Scientists have made various attempts to verify this experiment. More recently, they have even worked with a double-double-slit experiment. The result has remained the same: as soon as information becomes available about which route the particle has taken, its wave characteristics vanish. This has simply led to the conclusion thus far that light and electrons are both, particles and waves.

In other words, passing a double-slit puts an elementary particle (electron or photon) "into a state that matches a movement through both slits," but only as long as it is not being observed.[3] A rather humorous depiction of this appears in Fig. 12: we only observe the skier after s/he has passed the tree.

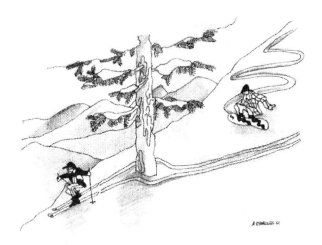

Fig. 12: The non-classical skier

As far as elementary particles are concerned, the measurement of the particle's characteristics (position) destroys its wave characteristics (interference). Strictly speaking, we should only speak of the path of a particle if and when it has been measured—between two measurements, it is not defined. The measurement itself is responsible for the fact that the sum total of all possible states turns into the concrete presence of one specific state.

---

[3] For the double-slit experiment, see Feynman (1999, 18-22); for the double-double-slit experiment, see Zeilinger (2005, ch. II.1). On the state of elementary particles passing a double-slit, see Schäfer (2004, 58).

The Atomic Model

The next development induced by the quantum hypothesis was the introduction of the atomic model by Niels Bohr (1885-1962). In 1911, he was working with Ernest Rutherford (1871-1937) in Manchester. Rutherford had just discovered that the atom consists of a tiny heavy nucleus which is surrounded by a cloud of orbiting electrons. His experiment, in which helium-nuclei are scattered on very thin gold foil, shows that most particles pass through the foil with only minimum deviation (see the black dots in Fig. 13). Under large scattering angles, however, significantly more particles were detected than should be expected according to Thomson's atomic model, which was developed in 1903 (see the dashed line in Fig. 13). The latter assumes a distribution of electrons in the atom comparable to that of raisins in a cake.

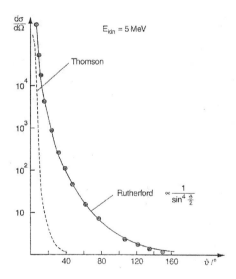

Fig. 13: Two atomic models compared to the experimental results

The increased backscattering, on the other hand, could be explained through the concentration of mass and positive charge in a point-like nucleus and the negative charge of electrons in an expanded shell (see the solid line in Fig. 13). The results of Rutherford's measurements thus led to the replacement of Thomson's model.

Next, Bohr asked how an atom could ever be stable. According to Maxwell's equations of electrodynamics, all negatively-charged electrons that circle around a positively-charged center, should emit energy and quickly (10^{-9} sec) drop into the nucleus — unless, that is, there are only very limited, permitted tracks along which electrons can move, including an innermost track which they cannot leave.

Based on the quantum hypothesis, Bohr developed an atomic model that explained the stability of atoms and the structure of the periodic table of elements. This had far-reaching consequences for our understanding of the chemical bindings of atoms and, in 1916, was expanded into the Bohr-Sommerfeld model. This says that the electrons move along tracks around the atomic nucleus (see the schematic representation in Fig. 14). Only radii are allowed whose angular momentum L of the electron is quantized according to $L = n\, \hbar$, with $\hbar = h/2\pi$ and $n = 1, 2, 3, \ldots$

The transition between two discrete tracks, that is, between two stationary states, occurs through the absorption or release of the difference in energy $\Delta E = h\, \nu$. In this manner "Bohr succeeded in connecting the strange stability of atoms . . . with Planck's quantum hypothesis — which is not really understood either . . ."

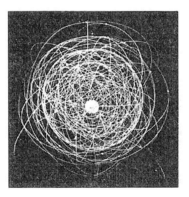

Fig. 14: Schematic representation of the uranium atom as seen according to Bohr with the nucleus in the center surrounded by the electrons.

Nowadays we no longer see electrons as orbiting along set tracks, but understand them instead as a standing probability wave which surrounds the nucleus. Transitions are possible, for example, from a state of higher energy to one of lower energy.[4] States of higher energy can be compared with higher harmonics, while the ground state is like the first harmonic. The approach of assigning a wave character to particles that have mass can be traced back to the Ph.D. thesis of Louis de Broglie (1892-1987): Planck's quantum of action h connects the momentum p (of a particle) with the wavelength λ (of a wave) by the equation $p = h / \lambda$.

The Formulation of Quantum Mechanics

Werner Heisenberg (1901-1976), a very young man in early 1925, succeeded in finding the first consistent formulation of quantum mechanics while convalescing after a bout of hay fever on the North Sea island of Helgoland.

His name is specifically linked to the uncertainty relation, which stipulates that the product of two conjugated variables (e.g., position and momentum; energy and time) in principle can never be determined more precisely than $\hbar/2$:

$$\Delta x \cdot \Delta p \geq \hbar / 2$$

with Δx as uncertainty of position

and Δp as uncertainty of momentum

This inequality says that we can no longer fully determine the state of a specific particle: position and momentum, which are observable parameters, have no simultaneous reality and can not be measured with high pre-

---

[4] For comparison of the atomic models, see Demtröder (2005, 68). On electrons in the atom, see Tegmark (2001, 70); Heisenberg (1969, 114). On the stability of atoms, see Heisenberg (1969, 47). For the uranium atom, see *Fischer Lexikon* (1981, 735).

cision at the same time. When Δx is small, and x is relatively precisely known, then Δp increases, i.e., p becomes relatively vague. The formalism of quantum mechanics thus only allows for statistical evidence. Still, it is possible to describe a situation in which a particle is approximately at a position Δx, i.e., is in fact a wavepack (Fig. 15), and has an approximate speed Δv = Δp / m (with m as mass). As long as the amount of inaccuracy remains small, there are no problems in the experiment. In that way, for example, the electron's movement can be related to its actually observed, much wider track in the cloud chamber.

In 1926, Erwin Schrödinger (1887-1961) succeeded in developing another consistent description of atomic processes. Since matter exhibits wave phenomena, he began his quest by searching for a wave equation. A quantum mechanical state is given by the wave function ψ, and Schrödinger's equation allowed the description of time evolution in quantum mechanical systems (see equation (1) below).

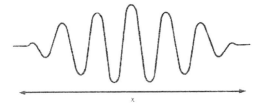

Fig. 15: An electron as a wavepack

Next, Max Born interpreted wave functions with the help of probabilities.[5] In doing so, he discovered that the absolute value of the wave functions squared $\psi\psi^*$ indicates the probability of finding the result within a certain range, such as, for example, the location on the screen where the electron lands. Since then, physicists have spoken about prob-

---

[5] On the first formulations of quantum mechanics, see Heisenberg (1925); Born (1926); Schrödinger (1926a). For details on the uncertainty relation, wave functions and probabilities, see also Audretsch (2002, ch. 1); Heisenberg (1969, 125); Dawydow (1972, 45).

abilities rather than certitude. Causality is no longer valid in the strong sense that entails determination of a result, as in classical mechanics; it now only indicates that event 2 depends in some way on event 1, i.e., statistically.

Let us clarify this once more with an example taken from everyday life: What happens when we drop a cup? In classical mechanics, it is clear that the cup or its pieces end up on the floor. In terms of quantum mechanics, there is a probability greater than zero that the cup could tunnel its way through the floor; this probability is very, very small, but it is not zero. The smaller and lighter the cup, the greater is the—still minute—probability of it ending up within or below the floor. As a consequence, the waves that are connected to particles are not "real" waves, but instead "probability waves." The act of carrying out the measurement defines reality.[6]

Schrödinger's equation on the time evolution $\delta\psi/\delta t$ of quantum mechanical systems:

(1) $i\hbar\, \delta\psi / \delta t = H\, \psi$ with H as Hamiltonian operator, which also describes the total energy of the system, and $i^2 = -1$.

When the total energy H is known, one can calculate the wave function ψ, on the basis of an initial value, at any time t.

Free particles can be described by a function representing a plane wave:

$\psi\,(r,t) = \psi\,(0)\, e^{i(kr-\omega t)}$ with position r, wave number k, and angular frequency ω.

This is a mathematical description that includes both the wave and the particle aspects ($k=p/\hbar$).

Schrödinger's wave mechanics defined the wave character of atomic processes and made them easier to calculate, while Heisenberg's matrix me-

---

[6] On "real" waves and "probability waves," see Capra (1988, 152); Zeilinger (2005, 146). On the comparison of both formulations, see Fischer (2004, 49); Schrödinger (1926b). For a general outline, see Feynman (1999, chs. 1-2).

chanics placed greater emphasis on their particle nature. It was Bohr who reconciled the two aspects with his concept of complementarity.

The Copenhagen Interpretation

Niels Bohr and Werner Heisenberg spearheaded the so-called Copenhagen Interpretation of quantum mechanics on the basis of the idea of complementarity, on the one hand, and of the uncertainty relation, on the other. They presented it in 1927 at the Solvay Congress in Brussels. Complementarity means that any experimental set-up that determines the particle characteristics excludes the simultaneous observation of the wave nature, and vice versa. There was, at the time, no experiment that allowed the observation of both characteristics at the same time. However, Bohr did not let himself be discouraged by these contradictions. He resolved them by immersing himself in the phenomena and developed a "true instinct" for useful assumptions about atomic events despite the difficulties. He succeeded by raising the enigmatic behavior of particles/waves to the status of a postulate.

Nowadays, the Copenhagen Interpretation is a widely accepted view of quantum mechanics. It is so called because many important concepts were developed in Copenhagen, notably at the Institute for Theoretical Physics (founded in 1921). Here Bohr realized his vision of a true scientific family: he had labs with scientists from many countries on the ground floor and lived with his family upstairs (Fischer 2004, 36). Known as the Socrates of natural sciences, he also seriously studied philosophy to go beyond the new physics and "contribute to the clarification of the preconditions of human knowledge."

For almost fifty years, he exerted a decisive influence on atomic physics, both as a human being and as a scientist. World War II, in turning scientific friends into political adversaries, destroyed this creative community. Nevertheless, Copenhagen remained an important center for numerous intellectuals who emigrated under the Third Reich. It was not until 1953, when the Centre Européen pour la Recherche Nucléaire (CERN) was founded in Geneva, that his vision of a European physics community was realized once again.

As Ernst Peter Fischer outlines (2004), Bohr loved company and often spent much of his free time with his colleagues. Once when staying in a ski hut with Heisenberg in 1933, he was on kitchen duty, and began pondering the connection between dishes and uncertainty, "We have dirty water and dirty towels and yet we manage to clean our plates and glasses with them."[7] This is the same as in language: only rarely do we reach truth directly with clear words; instead we typically approach a better understanding of nature with vague terms and limited logic.

As already pointed out in connection with the order of magnitude of Planck's constant or the quantum of action (10^{-34} Js), the effects of quantum physics are minimal in everyday life and mainly important on atomic scales—the atom, after all, is tiny: $\approx 10^{-10}$m, and its nucleus even more so: $\approx 10^{-15}$m. Still, examples from daily living can be very helpful to clarify the situation. For example, the nucleus of the atom in relation to the atom as a whole is about the same as the third of a grain of rice in a soccer field (1mm : 100m).

Schrödinger's Cat

So far we have focused on atomic particles. Why, then, should what applies to them not be equally relevant for all the objects in our normal experience? After all, macroscopic things consist of nothing but atoms.

Let us first look at this issue in light of a very practical ordinary life situation: We are taking a leisurely walk in the mall and would like to have an ice cream. Soon, we see a stand with an ice cream machine that has two levers: chocolate and vanilla. We order a cone but do not specify the flavor. Leaving the selection to the salesperson, we do not know which one we will receive. As long as the ice cream has not yet come out of the machine, both flavors are still possible. This means, the ice cream exists in a state of overlap between two states, which may lead to either chocolate

---

[7] For details on Bohr's way of thinking, working and living, see Fischer (2004, 8, 113, 102, 36); on Bohr's presentation in Brussels, his "instinct" and the connection between dishes and uncertainty, see Heisenberg (1969, 127, 62, 215).

or vanilla ice cream. The extrusion of the ice cream matches the so-called measurement process, in which one of the two possibilities becomes reality. One can further imagine a mixture of chocolate and vanilla—however, as far as this example is concerned, that particular lever is out of order.

To describe such a situation in quantum mechanical terms, let us now look at Erwin Schrödinger's famous thought experiment: A cat is in a closed steel chamber. Besides the cat, the chamber also holds a "hell machine," a radioactive source with a known half life. As soon as a single decay occurs, the Geiger counter reacts. Transmitted via relay, this signal causes a hammer to fall on a vial, which releases poisonous Prussic acid that kills the cat. Nobody knows when a decay occurs. The atom is thus in a state of overlap of decay/ no decay. For the cat, this situation means that nobody knows whether it is alive or dead. To describe its state after one half life in quantum mechanical terms, one has to sum up all possible states that the cat could possibly enter in the appropriate wave function ψ (see equation (2) below).

As equation (3) shows, there are additional mixed states besides the basic categories "alive" and "dead," which, however, cannot be observed. A quantum object occupies all its states at the same time—but only until it is being observed. In other words, things exist in a superposition of various possibilities, out of which one is realized when measured. Quantum mechanical characteristics are uncertain or undefined until someone asks about them and carries out the appropriate experiment. At this point, the wave function collapses, i.e., one state is realized and the situation or the observed state can be described in ordinary terms.

A different interpretation of this situation is Everett's model of many worlds. According to this interpretation, the cat is alive in one world, while it is dead in the other. The model also makes predictions that are consistent with the experiments.[8]

---

[8] On Schrödinger's cat, see Schrödinger (1935a, 812). On superposition of possiblilties, see Zeilinger (2005, 149-50); Laszlo (2005, 89); Audretsch (2002, ch. 1). On the model of many worlds, see Everett (1957).

The cat's state after one half life of the radioactive source is calculated by the superposition of all possible eigenstates of the system (f_1, f_2) that match the results "dead" = $|f_1|^2$ or "alive" = $|f_2|^2$:

(2) $\quad \psi = \Sigma_k c_k f_k$ with $k = 1, 2$ i.e., $\quad\quad \psi = c_1 f_1 + c_2 f_2$

The expectation value is derived from $\quad |\psi|^2 = \psi\,\psi^*$;

c_1, c_2 are scaling factors, characterized as: $|c_1|^2 + |c_2|^2 = 1$

(Audretsch 2002, ch. 1).

To determine the expectation value, i.e., the measurable values, multiplying and ordering yield

(3)

$$\psi\,\psi^* = (c_1 f_1 + c_2 f_2)(c_1^* f_1^* + c_2^* f_2^*) = |c_1 f_1|^2 + c_1 c_2^* f_1 f_2^* + c_1^* c_2 f_1^* f_2 + |c_2 f_2|^2$$

Besides the measured values "dead" and "alive," there are also unobservable mixed states (the terms with $f_1 f_2^*$ or $f_1^* f_2$), that is to say, "dead-alive" or "alive-dead."

Consequences

To sum up: 1. Instead of making exact predictions, we have to be content with statistic probabilities. 2. Atomic processes and measurement devices form one integrated unit. 3. The observer disturbs the system with the measurement. That means, our knowledge about the physical world is subject to intrinsic limitations that cannot be avoided, but can be predicted by quantum mechanics, notably through the concepts of uncertainty and complementarity.

Bohr's concept of complementarity joined these apparently irreconcilable opposites. In 1947, he received the Order of the Elephant from the Danish Academy of Sciences for his achievements. Interestingly enough, his coat of arms shows the Taiji symbol as well as the words "contraria sunt complementa" (opposites are complements) (see Fig. 16).

Niels Bohrs Wappen
Aus dem Gedenkbuch Niels Bohr, hrsg. von S. Rozental (North-Holland Publishing Company, Amsterdam 1967)

Fig. 16: Bohr's coat of arms

This brings us back from modern physics all the way to the oneness of opposites. Bohr's stroke of genius in recognizing and joining the contradictory opposites as complements, in other words, in uniting the opposites into a single Dao, solved the problem. Incidentally, he had traveled to China in 1937, when the Copenhagen Interpretation was already ten years old. I first saw the connection of wave-particle dualism with the yin-yang symbol in Capra's *Das Tao der Physik* (1988; *The Tao of Physics*). By putting the two together, Capra opened up a new perspective that stimulated public interest in modern physics.

This happened thirty-five years ago. Since then, experimental methods have greatly advanced: today we work with correlated systems. This has had tremendous effects, most notably on methods of information processing. Quantum physics forms the foundation for rapid technological progress, enabling an increasing number of applications of its phenomena. Besides lasers, microchips, and nuclear energy, developments are evolving in the direction of quantum cryptography, quantum teleportation, and quantum computer.[9] Even within philosophical postprocessing we have gone beyond wave-particle dualism and have moved to the next step, catching glimpses of the oneness that lies behind.

Quantum mechanics shows that there is only one reality: integrated oneness. It reunites subject and object, observer and measurement, separated radically by René Descartes 300 years ago in his division of the world into res cogitans (mental) and res extensa (extended). Their strange interconnectedness is the subject of the next experiment, which confirms the Copenhagen Interpretation as qualified.

The Einstein-Podolsky-Rosen (EPR) Paradox

Between 1927 and 1949, the many oddities of quantum mechanics led to an extensive debate between Einstein and Bohr. Albert Einstein was indignant about the idea that reality should depend on the process of measurement, pointing out decisively, "God does not play dice." He wanted to show, and David Bohm later expanded on this idea, that quantum mechanics is incomplete inasmuch as there must still be hidden variables to overcome its oddities and allow the recovery of the classical world view. This world view is known as the local realistic theory of nature; it says in Bell's words that "the result of a measurement on one system is unaffected by operations on a distant system" even if both systems interacted at an earlier time (1964, 195). Einstein's dissatisfaction with quantum mechanics is best expressed through the famous thought experiment that he developed in 1935 together with Boris Podolsky and Nathan Rosen. It marks the culmination of his debate with Niels Bohr.

---

[9] On correlated systems, see Aspect (1981; 1982a; 1982b); Tittel (1998). On quantum cryptography, quantum teleportation and quantum computers, see Bennett (1992); Bouwmeester (1997); Audretsch (2002, ch 6.2).

Only from 1972 onwards did it become possible to put his experiment to the test. Later, scientists succeeded in highly exact measurements on pairs of photons. Even later, Alain Aspect showed with time-varying analyzers that there could be no hidden variables since the experiments matched the predictions of contemporary quantum theory exactly.

The basic assumption in the original formulation of the EPR paradox is that two systems interact from a time t=0 up to a time t=T, and the wave functions of both systems before the interaction are known—i.e., particles A and B approach, collide, and separate again. Using Schrödinger's equation, one can calculate the states of the combined system at any later point in time. After the time T, the systems are separate. By measuring particle A at this point, one obtains the expectation value, for example, of the momentum. One could just as easily measure the location of the particles. However, this means that conjugated variables (position/momentum) would be measurable simultaneously, which would contradict the uncertainty principle.

Einstein considered this apparent contradiction to be proof of the incompleteness of quantum mechanics. Bohr's prompt response, however, made it quite clear that one could not look upon particles and measurement devices as separate, but instead needed to treat them as ONE system. In addition, a state is not defined until it is measured. Measuring the momentum precisely makes it impossible to measure the position. In addition, the physical question is only defined after the experiment is set up, i.e., after the decision has been made which parameter should be measured. Bohr thus refuted Einstein's objection. For quantum physical systems, one has to apply a completely different form of description than for classical ones—using the uncertainty relation and the principle of complementarity. This means that we have to completely rethink our most foundational assumptions about physical reality: locality and realism—the ideas that systems possess independent characteristics and that characteristics exist independently of a measurement—are completely "out." The new concept that takes their place is "entanglement."

In 1952, David Bohm formulated a variant of the EPR paradox for the spins of a proton system (see Fig. 17). Two protons emitted by a source Q are in a (singlet) state, such that the sum of their spins or the total spin is zero, and they are separated from one another. By measuring their spin along a predetermined axis, for example, the z-axis, one finds that there is a strong negative correlation between the spins. That is to say, if particle A has a spin of +1/2 (arrow up), particle B has a spin of -1/2 (arrow down). This is because the total spin of a system is a conserved quantity. The spin is one of the characteristics of particles (somewhat like eye color in people) and remains unchanged in a given system.

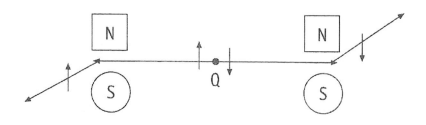

Fig. 17: D. Bohm's EPR experiment:

A source Q has two spin-½ -particles with antiparallel spin.

In a magnetic field (N-S) they align in opposite directions.

A total spin of zero and a measurement of particle A mean therefore that one can predict the spin of particle B. The observation, for example, of the z-component of one spin in a set of two particles thus immediately affects the other one. This is independent of how far the particles are separated from each other during the measurement. This seems to contradict the special theory of relativity, according to which information can be exchanged at the speed of light c at the very most. Depending on the experiment, impacts appeared even in systems whose distance could only be overcome with 20,000 times the speed of light.

Formulated according to Bohm, the paradox thus is that there is no explanation why particle B should have realized the same spin component (in this case, z) as particle A (with the opposite spin). David Bohm suspects that hidden variables are responsible for this effect. The connection with instantaneous influence between the particles is called nonlocal or entangled. Entanglement emerges from the superposition of (product-)

states of composed systems, with correlations developing as and when their various components interact. The term was originally coined by Schrödinger.

Nonlocality is a characteristic that we already encountered when discussing the collapse of the wave function at the instant of measurement. "What happens during measurement is that the state of the quantum system gets entangled with the state of the experimental device."

Entanglement has numerous aspects. It means, for example, that "maximum knowledge of a complete system . . . does not necessarily include maximum knowledge of all its parts." However, it does include a correlation between the parts. On the other hand, there is also entanglement in classical physics: assume you have got one pair of gloves and you are on the road and find that you have only the right one taken with you, then you know that the left one is at home. Seen from the perspective of quantum mechanics, it is quite obviously the case that particles which once formed a oneness no longer have an independent existence.[10]

This being so, it gets very interesting when looking at cosmic scales. Currently valid assumptions about the nature of the universe begin with its expansion after the Big Bang, which means that everything in the universe at some point was connected with everything else. As Fischer puts it, "the atomic reality . . . reveals a connection between individual objects that can only be described as wholeness" (2002, 189). Isolated particles without interaction that cannot be observed "are not part of physical reality." This means the universe is more a network than clockwork.

---

[10] On the debate between Bohr and Einstein, see Fischer (2004, 85, 90); Audretsch (2002, 12, 71). On locality and realism, see d'Espagnat (1979, 158); Audretsch (2002, 43). On the theoretical base of the EPR-paradox, see Einstein (1935); Bohr (1935); Bohm (1952a; 1952b, 1957); Bell (1964); Schäfer (2004, app. #15).

On the experimental results, see Freedman (1972); d'Espagnat (1979); Aspect (1981; 1982a; 1982b); Laszlo (2005, 99); Zeilinger (2005, ch. II.2); Ricard (2008, ch. 4). On the interpretation, see Heisenberg (1969, 193); Wheeler (1998); Fischer (2004, 87); Schäfer (2004, 58, 122, 265). On entanglement, see Schrödinger (1935b, 827; 1935c); Audretsch (2002, ch 8, ch. 9). For the mystic's comments, see Bingen (2001, 24-25, 249); Quint (2007, 326); Jäger (2005, 100); Ricard (2008, 101).

The core of the EPR Paradox can therefore be formulated as follows: two particles that have interacted at any time behave as ONE system, independent of how far the distance between them may be at a later point. This appears to be paradoxical only as long as one understands the two particles as independent entities. Once you see them as parts of ONE reality, the paradox is resolved, since there is nothing that has to be transmitted anywhere else.

Distances in space-time, it seems, do not separate quantum mechanical objects. That is to say, it is impossible to describe these phenomena as objective processes in space and time. Instead, it appears that space-time nonlocality is a key characteristic of reality, almost as if space-time did not exist at all.

Following Schäfer's interpretation, we should accordingly look for wholeness beyond space and time, on the level of pure consciousness. One such example is the sphere of our internal experience, which similarly is not part of ordinary space-time. Thus, as we enter a flow state in qigong, we have a feeling of timelessness; people entering mystical union lose all sense of time and space.

Thus Hildegard of Bingen says, "The light I see is not tied to space . . . As long as I envision it . . . I feel like a young girl and not like an old woman." Meister Eckhart similarly notes, "For the soul to recognize God, it has to move above space and time." If, therefore, we have access to a level of consciousness that experiences reality beyond space-time, and if reality is one integrated whole, does that mean that the universe as a whole has a consciousness? Does spirit or pure mind form the foundation of the universe and is matter "nothing but a dense version of this mind"?

Bell's Theorem

In 1964, at CERN in Geneva, John Bell worked on the EPR paradox from a mathematical perspective. He formulated group-theoretical considerations about possible correlations between different spin components in what was later known as Bell's Inequality:

$$n [A^+ B^+] \leq n [A^+ C^+] + n [B^+ C^+]$$

Here $n[A^+B^+]$ is the number of particle pairs with one component A^+ and another component B^+. The formula says that there is an upper limit for classical correlations between any two different spin components, that is, when locality and separability are also valid beyond classical physics (Audretsch 2002, ch. 6). The number of pairs with the characteristic A^+B^+ can be no greater than that of the sum of the pairs A^+C^+ and B^+C^+. The set up of Bell's inequality is explained in a very coherent way by d'Espagnat (1979). Now, quantum mechanics predicts that Bell's Inequality, due to the entanglement of systems, will not be valid for certain angles between the spin components.

This, then, is the meaning of Bell's Theorem: Under certain conditions the correlations between the spin components are different from what is permitted by Bell's Inequality (linear slope in Fig. 18). Experimental results fall very close to the curve as predicted by quantum mechanics. This means that quantum mechanics provides accurate predictions and presents a complete theory: no hidden variables are needed to explain the situation. Furthermore the difference from Bell's inequality manifests as a stronger correlation, just as predicted by quantum mechanics. Any assumption of hidden variables in quantum mechanics that lead back to a local realistic theory, would result in predictions that contradict quantum mechanics.

Bell's Theorem—i.e., the violation of Bell's Inequality—shows that there are no hidden parameters or valid local realistic theories. Still David Bohm's variant of the EPR-Experiment was a thought experiment, too. It was not until 1969 that physicists were able to devise experiments that could be carried out in a laboratory setting.

Beginning in 1972, the experiments were carried out, for the most part by indirect evidence, involving the polarization of photons, which is comparable to the spin of particles. To do so, for example, stimulated calcium atoms were used, which emit two photons in opposite directions during transition to ground state. The polarization of these photons can be measured and indeed turned out to be opposite. In other experiments, the polarization of X-rays that emerge during the annihilation of electron-positron pairs has been analyzed. They, too, showed opposite momentum and polarizations. Therefore, such experiments are comparable to Bohm's variant.

It has thus become obvious that these instantaneous effects between the different components of a system do in fact exist. There is, as noted earlier, a conflict with the special theory of relativity and its claim that no information exchange can happen faster than c, the speed of light. However, it can be shown that no information is exchanged, thus eliminating this conflict. It can be assumed that the correlations are not between particles or physical states but only between quantum states, that is to say, between the distributions of probabilities which determine how the particles

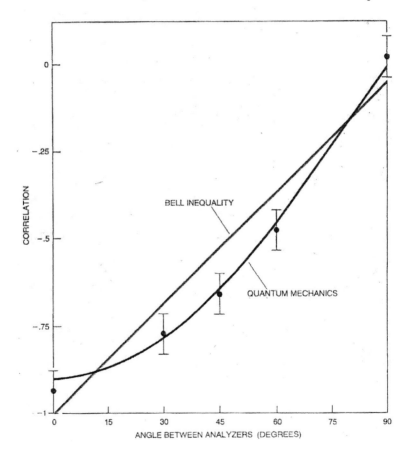

Fig. 18: Experimental test of Bell's Inequality

behave.

Quantum mechanical states are mathematical parameters in an abstract space, the so-called Hilbert space. The wave function this describes is a tool for the description of probabilities or probability waves. It should not be seen as a description of real waves, lest one gets entangled in contradictions. To make them observable, a measurement must take place. This, then, shows that "the behavior of real particles can be understood as and when they are described through wave functions which themselves do not match any reality." Obviously, this led to a great deal of discussion.[11] The vigorous debate between Bohr and Einstein was, after all, also a debate about ultimate truths and about God or, we might say, about Dao.

This Dao that we have now arrived at is, of course, a very special Dao: the polarity between reality and nonreality. Atoms and other quantum objects are no longer things but "observable phenomena" we perceive and which only exist in relation to us—in a world of mere probabilities and possibilities. Or, as Heisenberg formulated it, they are "between the idea of a thing and an actual thing . . . ."

Another special Dao appears in the wave-particle dualism that Bohr describes in terms of complementarity, where the process of measurement itself ends the possibility of further transformation. This is as if we did not know whether it was day or night until we checked the position of the sun and then were stuck with that position as it was. Critics even called the Copenhagen Interpretation the "big fog from the north" or "atomysticism."

Quantum mechanics has thus come to face a similar situation to mysticism: there is a clear understanding of a certain level of reality, but the only way to explain or speak about it is in images or parables, respectively in terms of operators and expectation values. Bohr even found this reminiscent of the "wisdom of the Chinese," who express their wisdom in anecdotes. In this context, he may have thought of the *Yijing* which the writer Hermann Hesse called "a system of parables for the whole world,"

---

[11] For Bell's inequality, see Bell (1964); on the experimental realization, see Freedman (1972); Aspect (1981; 1982a; 1982b); d'Espagnat (1979). On the conflict with the special theory of relativity, see Audretsch (2002, 53); Fischer (2004, 83-93).

or which Zhuangzi described as, "Nine tenth of my words are parables: that means I use external images to express my thoughts" (ch. 27).

To the present day, physicists do not understand how to explain the collapse of the wave function at the instant of measurement. There is no equation to describe it, but there is an alternative interpretation of quantum mechanics, which sees the mutual interaction between object and environment as its cause. Since quantum mechanics represents the more encompassing system, it must include the description of the previously valid system and show how the transition from quantum to classical physics works. That is to say, despite the fact that it was developed on the atomic level, it must be equally valid for the macrocosm.

The central point seems to be the phenomenon of decoherence, which means the destruction of the interferences by mutual interaction with the environment.[12] We already encountered this in connection with the double-slit experiment: when undistinguishable paths (which slit used?) are made distinguishable (observation), interference vanishes (only particle characteristics remain). Macroscopic systems cannot, obviously, be isolated from their environment to the degree that they are completely free from mutual interaction. Apparently, the environment has the same effect on macroscopic objects as measurement does on the atomic scales. This leads to combined wave functions and entanglement, and only the system as a whole is coherent.

If we now measure quantities of a partial system, we again observe classical values because these are states unaffected by decoherence. This means, decoherence is the key factor that leads to the observation of classical characteristics: it is a direct consequence of nonlocality. All other correlations are imperceptible. According to this, the macroscopic characteristics of the system are not intrinsic, as Aristotle thought (Metaphysik, Buch VII), but emerge through interaction with the environment. Indeed, the effect of decoherence is exactly the same as the collapse of the wave function. Bohr's method of "empathizing and guessing" guided him well in the context of the Copenhagen Interpretation (Heisenberg 1969, 64).

---

[12] On quantum objects and their reality, see Ricard (2008, ch. 5); Schäfer (2004, 100); Heisenberg (1969, 47, 326); Tegmark (2001, 72). On decoherence, see Zeh (1970); Scully (1991); Zurek (1991); Tegmark (2001).

Multiple Dimensions

Reflecting on how minute the fragment of reality is that we can perceive at all, we should not be surprised that we are out of our league when faced with microscopic phenomena. What is unusual is that we are at all capable of having these thoughts: should the subtle harmony of natural constants be even minutely different, we would not even exist, let alone be able to think. Unlike the n-dimensional Hilbert space in which we can perform our calculations, we can just about orient ourselves in three dimensions and for the very short period of our earthly life—at least in comparison to the age of the universe. We find it hard to even accept time as a fourth dimension—whatever goes beyond that is incomprehensible and unimaginable. The *Zhuangzi* even notes that "life is finite, knowledge is infinite. To pursue the infinite with the finite is fraught with danger"(ch. 3). The complexity is best illustrated with an example: let us imagine we are restricted to only two dimensions and see how life feels from the perspective of a two-dimensional being.

First imagine a surface with two designs: a circle and a rectangle. They seem to be two completely different things (Fig. 19). By looking at them in three dimensions, however, we see that they originate from the same object: a cylinder and its projections along two perpendicular planes.

In other words, there is only ONE three-dimensional object which, due to different projections onto the two-dimensional plane, appears as two different things. Could it be that particle and wave are merely projections of a higher-dimensional reality onto our three-dimensional world? When differentiating between particles and waves, we measure two different entities. By using quantum mechanics, we know that they belong together, but do not understand exactly how. The concepts of complementarity and decoherence are possibilities for dealing with this phenomenon. But where do these considerations of multiple dimensions lead?

From architecture we know the distinction between two- and three-dimensional perspectives: three two-dimensional projections can represent a three-dimensional object. For example, to build a house, we need a floor plan, a side view, and a front view. Could it similarly be possible that electrons and waves are merely three-dimensional projections of a four-dimensional reality? What might the four-dimensional reality look like that lies behind them? In any case, we would then need at least four three-dimensional projections to open access to a four-dimensional object. So far, however, we only have particles and waves. Are there other candidates in the scientific repository of oddities? Is it, perhaps, even the case that we need ten dimensions of space and one of time, as required in the string theory, to develop a complete theory of reality—a theory that, moreover, would connect the general theory of relativity and quantum physics? Why we do not notice anything from the other dimensions is very nicely worked out in Greene (2000, ch. 8).

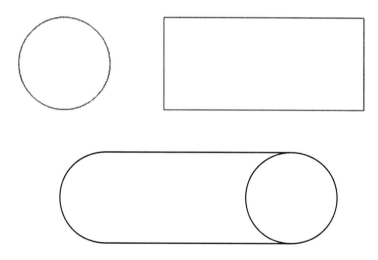

Fig. 19: Circle and rectangle as projections of a cylinder

All manipulations we apply to a flat surface from a three-dimensional perspective are entirely unintelligible to a being that exists in only two dimensions; for us, as three-dimensional beings, they are perfectly natural. Now imagine a three-dimensional (3D) being stepping on a flat surface (Fig. 20). The two-dimensional (2D) flat beings who live on this surface, only notice the outline of the other being's foot. When the 3D being moves on and lifts its foot off the surface, the outline of the foot vanishes. For 2D beings, this is entirely unknowable: quite likely, they make every effort to look for explanations and try to uncover natural laws that might explain this strange phenomenon—a phenomenon that is totally clear for any 3D being. Just think about it: could it be that there are so many unexplained phenomena in our world because our dimensionality is so limited? Will the contradiction dissolve by itself once we move into four or more dimensions of space?

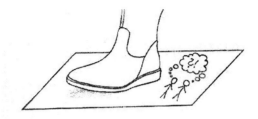

Fig. 20: A three-dimensional being appears in a two-dimensional world

At this point, we return to the question in chapter 3 of whether quantum physics forms a Dao of its own. The answer is yes. One reason is the well-known wave-particle dualism, but more important is the phenomenon of entanglement. Reality is much more strongly correlated than we could ever have dreamed of.

An earlier question concerned the mutual transformation of polarities. This has become much clearer since the experimental possibilities allowed single photon sources:[13] one photon leaves the source and moves through the double-slit like a wave, and it generates a single point of light on the screen, at a position that matches the interference pattern. This means that the photon arrives like a particle but, having the probability of default of a wave, arranges itself into an interference pattern. This is about as crazy as it gets!

---

[13] On single photon experiments, see Scully (1991); Audretsch (2002, 65); Belzig (2008).

All this suggests that the earlier thesis holds true: "To understand the connections in the universe, it is essential to create a oneness of the natural sciences and mysticism and move away from seeing these two as opposites."

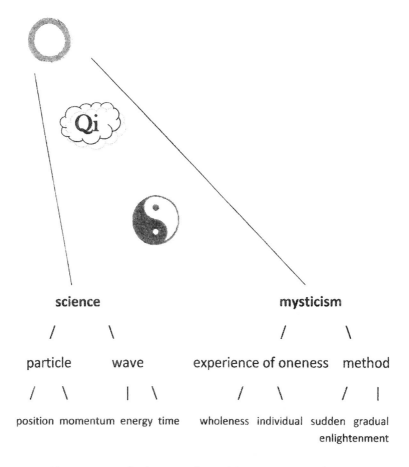

Fig. 21: Natural sciences and mysticism seen as complementary

Fig. 21 shows schematically the structure analogous to the splitting of polarities which continues unabated. The scheme starts with Dao, separating under the impact of qi into yin and yang, that is science (yang) and mysticism (yin). Science (here quantum mechanics) is again a Dao that expresses itself in particles (yang) and waves (yin). Particles, then, can be described by the complementary factors position and momentum; waves, by energy and time. The right side of the figure then shows mysticism as one Dao, with its yin (experience of oneness) and yang (methods) aspects. They, in turn, have further branches. Mystical union includes the complementary aspects of belonging to oneness, and yet, being individual and separate. The Method category similarly involves both polarities, such as, for example, sudden and gradual enlightenment in Zen.

Here is another good place to stop and allow these new ideas to sink in. To do so, take a break and enjoy the practice, working preferably with the first three exercises of the Second Cycle, "Transformations" (see ch. 7 below).

6. Synthesis
Pulling It All Together

When I woke up,

The sun was coloring the sky,

The moon took along the mist

And vanished – in mystery.

These various issues and phenomena continued to percolate in my thoughts and practice over many years. The various pieces of the puzzle only came together in a cohesive picture when I realized that the background theme of all of them was the fundamental human yearning for wholeness (Heisenberg 1974, ch. 3). Some believe that human beings are driven most of all by the search for knowledge or the quest for wisdom. However, it is quite obvious that true human yearning goes far beyond that. I believe that it is essential for our inner vitality to give ourselves permission to feel this fundamental yearning and not suppress or ignore it. Whether or not we can ever satisfy it, may well be of minor importance.

In the beginning of his *Metaphysics*, Aristotle says, "All human beings naturally have a yearning for knowledge" and, I would like to add, for meaning. This is the basis of much human exploration and also the root of my question about the true nature of reality. St. Thomas of Aquinas (d.

1274) speaks of this basis as follows, "Human beings are naturally equipped with the desire to experience the truth of God."[1]

In pursuit of the structure of reality, we looked at three completely different areas: qigong, mysticism, and quantum physics. The result was that all three have the same basic theme, which they call by different names: Dao, oneness, and entanglement.

In qigong, we practice approaching the completeness of being following the changes of yin and yang by opening and closing, rising and descending. We return to Dao by smoothly connecting the complementary phases of each exercise.

In mysticism, we encounter wholeness in stillness and in the experience of complete, universal oneness. By entering a transpersonal level of experience, we overcome all worldly duality for a period.

In the natural sciences, we work with entanglement or nonlocality, as first described in the EPR paradox and the violation of Bell's inequality. It shows an entirely unexpected oneness between the subject and object of measurement, respectively, between once connected parts of a system.

Does entanglement show us in our experiments a higher reality along the lines of what mystics become aware of in their transcendental states? Is it true that in the center of being there are no material particles but pure cosmic consciousness? Heisenberg notes, "The first draft taken from the grail of natural sciences may make you an atheist—on the bottom of the grail, however, God awaits." He believed that one can connect at any time to the "central order of things."

The wave function forms the bridge between things and the Dao. It describes the sum of all states and is valid without distinction for particles and waves. The diligent separation of subject and object, long practiced in the natural sciences, is now making way for a closely interconnected reality. What would happen if professors revealed this truth to graduate students of physics already in the first semesters, instead of waiting for them to discover it for themselves as they mature? Internal and external experiences run together and complement each other: there is an integral coher-

---

[1] On the yearning for knowledge, see Aristoteles (2007, 37); on St. Thomas of Aquinas, see Kopp (1997, 99).

ence between scientific and mystical understanding. This becomes tangible in qigong—even in the small segment of a single exercise (Bock-Möbius 2010c).

Heisenberg says, "I think that the effort to overcome opposites, to reach a synthesis of rational understanding and mystical union is the foundational myth of our age—it is what we all pursue, whether consciously or unconsciously." I personally would replace the word myth with goal: it matches exactly what I propose in this book. My core understanding is: "The synthesis of rational understanding and mystical union is the goal that we all pursue today, as well as being our most essential yearning."[2]

Heisenberg also notes that, historically, the most fruitful developments occurred whenever two different forms of thinking met. Does this also hold true for science and mysticism? I should think so. If not, we should assume that there are no connecting points between object and subject, knowledge and experience, mind and matter. Both are basically just different ways of understanding or perceiving of reality: an outer and an inner one. Together they provide an apperception of reality, not only through the brain, but also "through the heart" (Heisenberg 1969, 52). Such an understanding has nothing to do with romanticism and everything to do with wholeness.

We can verify a hypothesis just as easily through internal experiences as through external experiments without giving up validity. It is not necessary to always measure things. In his book *The Natural Sciences and Religion*, Ken Wilber notes that the empirical method at the foundation of science applies equally to outer as well as to inner experiences. In the old days, the three realms of ethics, science, and art were not separate. This had the disadvantage that religion dominated both science and art, leading to negative situations as, for example, in the cases of Galilei (1564-1642) and Michelangelo (1475-1564).

---

[2] On transcendental states and cosmic consciousness, see Jäger (2005, 138),; Schäfer (2004, 10); on the "grail of natural sciences" and the cosmic order, see Jäger (2000, 102); Heisenberg (1969, 332-33). For the "myth of our age," see Heisenberg (1974, ch. 3).

Differentiation in itself was welcome. However, this differentiation among the three realms in the modern age has developed into a radical separation. This has led to scientism insisting that there is no reality outside of science, thereby denying the existence of all internal dimensions and reducing them to externals. In other words, the development has become too strong for its own good, overshooting its original goals and leading to a world without meaning.

On the other hand, science has led to some truly extraordinary feats. One can, for example, use measurements of specific brain waves (EEG) to make detailed statements about physical and mental relaxation. Still, science has no systematic and pervasive means to explore internal structures: it pursues truth where religion seeks meaning.

This has to be kept in mind. Its main operative model, the tripartite empirical method (hypothesis—experiment—falsification) can just as easily be applied to mental and spiritual experiences. It can supply criteria that help us examine whether a certain internal experience provides a particular way of understanding the world and thus makes a real statement about the cosmos. Nevertheless, the reconciliation of science and religion cannot be a one-sided effort. It must also be supported by the other side. In this context, Ken Wilber believes it is essential for religion to recover its central concern of mystical experience and create very clear boundaries between those aspects of its beliefs that are of a more mythological nature.

The fundamental assumptions in this endeavor are that the world is ordered in an intelligible manner and that the same structures apply both internally and externally. In looking at the current progress in modern physics, it becomes evident that these assumptions are justified: more and more phenomena that used to be seen as isolated and separate are now accepted as interconnected. Already in the seventeenth century, Isaac Newton connected the sky and the earth through the laws of gravity, valid equally among planetary systems as under the apple tree. In the nineteenth century, James Clerk Maxwell (1831-1879) showed that electricity and magnetism are just different expressions of the same natural force. In the twentieth century, Albert Einstein linked space and time in his theory of relativity. Today, in the twenty-first century, physicists are working on the unification of the four fundamental interactions—electromagnetism, weak force, strong force, and gravity—with the goal of reducing them to one coherent power.

The astrophysicist Trinh Xuan Thuan says, "Quantum mechanics is making huge progess in the calculation of phenomena, but is hardly moving at all when it comes to philosophical foundations."[3] By connecting the considerations related to oneness and entanglement in quantum physics (ch. 5) with the visions and applications of qigong and mysticism (chs. 3, 4), this might change. The concept of nonlocality opens the gate to transcendental parts of physical reality that go beyond the limits of space and time—aspects that cannot be perceived with our ordinary senses and their technical expansion in the measurement devices. Nonlocality is a form of behavior that denies or overcomes all temporal and spatial dimensions, "The mystery knows neither space nor time" (Swami Sudip 2004).[4]

Schäfer also presents a very interesting position about evolution. He notes that modern biology still behaves as if quantum mechanics never happened. The molecular processes leading to mutations are quantum processes. Their random occurrence stays within a certain preset frame, determined by the possibility of quantum states. This means, mutation is the occupation of one of these virtual quantum states, thus making it actual (2004, app. 18). From this perspective, Schäfer sees the universe as a network of nonlocal nodes whose transcendent order is stored virtually, then projected into matter. He thinks that the mental or spiritual order of the universe is mirrored in people as ideals. Our spiritual needs thus echo the consciousnesslike background of the universe: our ability to perceive these needs shows that human consciousness is linked with this background.

Oneness as a foundational principle appearing in all these different realms indicates that it is of exceptional importance. Furthermore, the idea of oneness is also essential in numerous other areas as, for example, in

---

[3] For the connection between mind and matter, see Ricard (2008, 38, 405). On EEG during mental relaxation, see Haffelder (2006, ch. 1). On the application of the empirical method to internal experiences and a meaningful world, see Wilber (1998, ch. 12, ch. 15); Heisenberg (1969, 160). On the connection of phenomena and on philosophical foundations, see Ricard (2008, 76, 162-63);

[4] On the limits of space and time, see Schäfer (2004, 231).

Buddhism with its focus on the correct vision and apperception of reality, as documented persuasively in *Quantum and Lotus* by Ricard and Thuan (2008, ch. 1, ch. 4). The belief is that reality can only unfold in the mutual dependency of agents and that our world consists of a network of relationships created through the interaction of consciousness and phenomena. In actual fact, all apparent phenomena are empty. "The oneness of emptiness and the world of apparent phenomena [is] the last secret on the path to understanding reality" (2008, 235). Things only exist as a manifestation of the divine. Buddha, the Enlightened One, came to realize the indivisibility of the world and was conscious of the integral wholeness of all apparent phenomena.

It is also a core notion in Buddhism that the true nature of things is beyond all duality. At some point, the human mind lost sight of the original oneness of emptiness and apparent phenomena, instead creating an artificial boundary between consciousness and the world and separating self and not-self (2008, 59-60). This is why we believe that things are solid and permanent, and that our self and personality do in fact exist. In actual fact, our self is nothing but a mirror image—entirely dependent on the observer, "We are nothing but a swirl in the river" (2008, 188). The *Yijing* says similarly, "All flows along like a river, without stopping, day and night." This is echoed by the ancient Greek thinker Heraclitus (ca. 535-475 BCE), "We cannot step twice into the same river."[5] Our ignorance regarding the true nature of reality, according to Buddhism, is the cause of all suffering, producing our neverending cravings and desires (Ricard 2008, 54). The best remedy is to understand it correctly and embody this understanding fully. This means that Buddhism, too, has a Dao: emptiness and apparent phenomena.

By beginning with the assumption that all things only exist because they arise together with their complementary opposite, we recognize a direct physical equivalent in vacuum polarizations. In a space where there is essentially nothing—in a vaccuum—particle-antiparticle pairs can arise for very short periods of time, as long as they are consistent with the uncertainty relation. Examples include electron-positron pairs that duly enter

---

[5] On the volatility of phenomena, see Wilhelm (1950, intr.); Störig (1985, 135). On the mutable world, see Ricard (2008, 84, 149).

into further reactions and interactions. Particles and antiparticles together form yet another Dao.

There is a brief moment of creation: . . . – nothingnesss – yin-yang – nothingness – … and so on and so forth. Can we transfer this into the realm of cosmology: . . . – nothingness – big bang with expansion and contraction – nothingness – . . .? This would bring us to the cyclical universe and the cosmological mode of Buddhism, where arising and passing away never ceases. Buddhists—as much as early Daoists according to Zhuangzi—never bother to define an original first cause: anything that contains the first cause in itself has to be immutable and cannot interact with anything else, which means it is also unable to create a mutable world. Thus Zhuangzi says:

> There is a beginning. There was a beginning before that beginning. There was a beginning before that beginning before there was the beginning. There is existence. There was non-existence before that. There was existence before the beginning of that non-existence. There was non-existence before the existence before non-existence. (ch. 2)

The solution Buddhism offers with regard to the particle-wave dualism is again the emptiness of things. A Tibetan poem says, "Emptiness is not empty of function. It is empty of reality, of any form of absolute existence." Since apparent phenomena have no existence of their own, and are therefore an unreal entity, they can manifest at times as particles and at times as waves. At this point, at least the way I see it, Daoism goes another step further. Not only does the unfolding occur in mutual interdependence, but all unfolding happens in complementary pairs. This, I think, comes even closer to the nature of reality.

"The goal of science is to expand and order our experiential knowledge. It does not aim for an explanation of the true essence of phenomena," Niels Bohr states. That is, science wants to make predictions about the experimental behavior of phenomena. Audretsch has a somewhat different take on this. He thinks that it is "one of the tasks of a physics theory to go beyond predicting the results of measurements and lead to general conceptions of the physical world" (2002, ch. 1). In other words, interpretations are still part of physics, even if they have metaphysical implications.

Science has never made it its goal to show how to achieve peace and happiness. However, as human beings we want to know why we suffer and how to become happy. Obviously, this has a great deal to do with our yearning for oneness. Of course, we are glad to know about scientific findings, but not when they lead to inhumane applications as, for example, in nuclear weapons, and when they may lead into an uncertain and potentially hazardous future as, for instance, through genetic engineering. Mystical spirituality would thus be an ideal partner for the natural sciences: it would keep things focused on the human plane, prevent excesses, and assist in the true realization of wholeness.

"The key to the Bell Inequality is that it expands the dilemma we face due to the quantum phenomena into the macroscopic world."[6] With some modifications, entanglement is already being applied in practice: although we don't yet have the quantum computer, quantum cryptography is already being used for subtle codifications. This, of course, does not prevent decoding but it allows the operators to know if interception occurred.

Modern life has a tendency to remove us far from nature. Through qigong, art, and other methods of cultivation we recover nature and return to Dao. All these paths "must lead into the mystery, into another dimension, lest they remain mere techniques" (Swami Sudip 2004). The oldest form of the mystical experience is probably becoming one with nature. It is an experience that goes beyond anything that human beings can perceive. I think that we as human beings begin to suffer as soon as we no longer feel connected to this oneness.

There is no such thing as an isolated particle. Can there be a completely isolated human being? Human beings are social beings: they only feel comfortable as part of a social community. Is, then, our true social community not just our immediate environment, but possibly the entire universe? The truly astonishing fact is that it is not only a few isolated, weird mystics who speak of experiences of union nor certain possibly detached qigong practitioners. Instead, hard-core, mainstream, and (almost) normal natural sciences show conclusively just how tightly even the most

---

[6] For the following quotations, see Ricard: Tibetan poem (2008, 96); manifestation as wave or particle (2008, 120); Bohr's statement on the goal of sciences (2008, 405); on the key to the Bell Inequality (2008, 128).

distant phenomena are interconnected. This means that we all carry cosmic, universal responsibility in our lives.

Our knowledge that consciousness and the world of phenomena are not separate is one thing. To actually experience this in body and mind, to intimately feel just how interwoven mind and the outside world truly are, is another. It is from this position that a fundamentally ethical attitude toward all life, "compassion for all sentient beings arises." We cannot perceive the true nature of reality, but only experience it, precisely because it goes beyond all senses and thinking.

The conclusion I draw from this is that we begin to recover wholeness as and when we connect to the oneness beyond the polarities, and as we manage to overcome the duality between body and mind.[7] This can be perceived directly in qigong practice, which can be understood as a subtle approach to the transpersonal state. A very good possibility in this context is to practice qigong with the specific intention of linking with the movements of Heaven and Earth. As we expand our intention to include a multi-level and multi-disciplinary universal context, our understanding becomes more inclusive and our practice deepens. This, in turn, may well have a positive effect on our health and harmony.

Since each qigong exercise overcomes the polarities, for me, any qigong exercise is a minimal form of oneness: it is a quantum of unity, a ladle-full of Dao (Bock-Möbius, 2010a). It is not possible to condense a qigong exercise beyond the basics of opening and closing. The harmony of the interrelationship between the polarities in each exercise is the essential in every action. Being conscious of practicing the recovery of Dao, which manifests in so many different ways, we can—as I see it—clearly enhance the internal power of practice.

When I teach qigong to 9th and 10th graders, I find myself on occasion confronted with interesting questions like, "What does waving our arms about have to do with staying healthy?" The answer is: because we connect to the inherent nature of things, because we are one with the universal rhythms. Carried by universal meaning, we are "waving our arms

---

[7] On interception, see Röthlein (2004, 102); on becoming one with nature, see Borchert (1997, 43). On the ethical attitude and on the nature of reality, see Ricard (2008, 28, 53).

about in harmony with Heaven and Earth." This brings us in harmony with ourselves. However, before we begin the actual practice, all those thoughts should be left behind. Instead, we should come to a state of stillness and go back to Jiao Guorui's first key point: naturalness. From here, we can enjoy our practice without concerns.

At this point it is useful to recall the quotation from Wu Zhen that was mentioned in the Introduction, "Having fully realized the cosmic principle that rests in emptiness—what sorrows could still fill the heart?" How is that? By coming to the "pivotal center of meaning," as Zhuangzi says (ch. 2) and by perceiving this meaning, we come to understand the laws of nature (ch. 17). We understand the inherent process of nature, as much as the connections between humanity and nature. Having understood these experientially, we are no longer "pulled along by mere phenomena" (ch. 5).

As soon as we manage to "behave like the perfected and no longer force anything even if we could force it, we remain free from excitement" (ch. 27). This allows us to lead a life that does "no harm" (ch. 17). We no longer need to worry about theoretical hair splitting, being instead able to fully concentrate our mind on eternity and stillness (ch. 27). This means we are coming closer to the state of the "true person" or "perfected one," who maintains a "perfect balance of being both natural and human"(ch. 6), and lives in complete balance of yin and yang, his *qi* flowing harmoniously, and his channels open (Jiao 1993a, #106).

The core understanding here is the stillness within the Dao. This means, as Zhuangzi expressed it already, "to reach the central experience . . . that lies beyond thinking and is only partially appreciated by science" (Wilhelm 2006, 26). It is, as Jäger says, the attainment of transpersonal consciousness that contains all healing powers (2005, 100). "Whoever manages to leave his self behind will perceive the natural so being of all things," Zhuangzi notes (Wilhelm 2006, 35). This sounds much like the Buddhist vision, "The personal experience founded in meditation leads most of all to the understanding that the personal self . . . has no real foundation."

Why do we have to think so much when we could just stick to plain facts? Einstein says, "It is wrong in principle to try and reduce a theory solely to observable factors, because in reality the opposite is true: the theory decides what we can observe" (Audretsch 2002, 190). This means, the more encompassing we create our theory, the more we can observe. Be careful! If we only rely on theory, we eventually know everything except the central core. The mere accumulation of knowledge leads nowhere. Personal experience alone refines our knowledge into wisdom, not unlike in an alchemical process that runs in the background of our lives and makes us act responsibly. In qigong, we start to become conscious of this underlying process and learn how to support and enhance it.

The experience of our consciousness suggests "the existence of a continuum of consciousness that cannot be reduced to any material foundation." Quite possibly, there is a parallel continuum of matter and energy. In Buddhism, the world and consciousness have always been closely linked—like the two halves of a walnut. To get closer to oneness, we don't have to do very much, "the simple process of breathing already connects us to creation."[8] Here breathing is considered as opening and closing, being a single qigong exercise that gives us a quantum of Dao, a quantum of oneness.

It all being so natural and obvious, we have essentially already arrived. Why, then, are we still seeking? Because we live in a polarized world and "our brain loves explanations," as Swami Sudip outlines (2004)—even those explanations that point out its own limitations.

> *At first creation, induced by qi, Dao separated into yin and yang;*
>
> *At any moment, practicing qigong, yin and yang rejoin in oneness.*

Thus the circle closes. We realize that all our striving in the various fields of expertise reflects the same underlying cause. In the words Su Zhe places in the mouth of the inspired painter Wen Tong with regard to the secret of his art:

> *Dao is what I love.*

---

[8] On the foundation of personal self, on the "central core," on "continuum of consciousness," and on the "process of breathing," see Ricard (2008, 39, 394, 281, 402).

To experience this understanding in practice, let us now move on to the fourth exercise of the Second Cycle, "Transformations" below: "The Five Wheels," followed by the concluding exercise, "Closure," which together allow a smooth transition back to everyday life.

Fig. 22: The Huangshan Mountains with typical pine trees

7. Practical Application in Thematic Cycles

„… and it is the eternal One,

manifesting in so many ways …"

Goethe: *Parabase*

There are many varied, even innumerable forms of qigong practices, often called by colorful names. Still, they all have certain core commonalities, clearly recognizable among different styles and the modes of various teachers. For this volume, I have selected practice sequences that are particularly suited to its theoretical exposition, that is, they help to open or intensify our experience of the oneness of polarities.

Preparation

1. Beginning Position

Stand up straight with feet parallel, toes slightly turned out. Pivot on the right heel to turn the foot forward, then shift the weight to the right and step the left foot to the side. Feet are shoulder-width apart and parallel.

Next, squat slightly. Gently bend the knees as if sitting on a cloud or the edge of a table. Shoulders are relaxed, arms hang down loosely, head rests lightly on the shoulders, with eyes looking straight forward. Breathe naturally, maintaining a firm and secure posture.

Now, let your attention sink into the elixir field in the abdomen. Intensify or release the sitting posture until you find the position that is best for you. It should support both a straight spine and a relaxed back. Imagine an inner closing in hips and pelvis as though collecting something or an inward-turning of attention (Jiao 1989, 35). As you maintain attention, your breathing naturally deepens.

Open your internal space to the image of standing just like a tree, but only to the degree that you remain comfortable. Jiao Guorui calls this practice "Standing like a Pine Tree," which not only implies deep roots but also resilience and self-sufficiency (1989, 31). Pine trees grow in the most-limited of conditions, rising up even from cracks in the rocks. People especially recognize the pines of Huangshan 黄山 due to their unique shape, character, and *qi* (1996c, 37).[1]

2. Standing like a Tree

Standing as described above, project the image of a tree from your elixir field into your whole body (#1). Imagine roots growing from your feet deep into the soil, providing support and stability as well as a connection to the earth. Thus well centered and thoroughly anchored, feel a light wind rising from your waist that causes your elbows to move slightly away from the torso. Imagine there are balls of air under your armpits that support your arms and shoulders. Your fingertips are a few inches away from your clothes.

Imagine your legs and torso as the trunk of the tree, your arms as its branches, and your head as its crown. Connect upwards to Heaven, the yang pole, while linking downwards through the roots with Earth, the yin pole. Stand like a tree between Heaven and Earth. Maintain this posture for several minutes.

---

[1] I call this exercise "Standing like a Tree," since I want to leave open whether the tree has needles or leaves. For me, it tends to be leafy, which may well have to do with the fact that in Germany we tend to think of oaks as the archetypal tree, firm and stable, see Bock-Möbius (1993, 44).

1: Standing like a tree

2a: Pressing two balls . . .

2b: . . . into the water

3a: Massaging the organs - vertical

3. Pressing Two Balls into the Water

Relax the wrists and move your fingers to hip level, tracing a horizontal circle and moving outward and away from the body, as if painting the Taiji symbol (#2a). Allow the air you enclose with this hand movement to condense into balls of *qi*. Press them gently onto an imaginary body of water. Doing so, allow your sitting posture to change by softly raising and descending the pelvis. End this with the hands placed sideways, thumbs near the seam of your pants, fingertips slightly turned toward the center of the body (#2b). This posture is also a good starting point for an internal massage; it is especially beneficial for ailments of the upper abdomen.

4. Massaging the Organs

Press the air ball under your left hand a bit more strongly onto the water, letting it reach the groin area. While doing so, allow the right ball, in a slow and controlled movement, to rise to about the level of the navel (#3a). Vice versa, press the right ball a bit more strongly onto the water and allow the left ball to rise to navel level. Repeat once, twice, or three times.

Bring the hands back to the same level. Roll the ball in the left hand slightly further to the back, while allowing the one in the right hand to roll further forward (#3b). Vice versa, let the left hand move forward while bringing the right hand back. Repeat once, twice, or three times.

Return both hands to the same level. To enter the third part of this exercise, move the hands in horizontal circles: the ball in the left hand moves back a quarter turn, the one in the right hand moves forward a quarter turn (#3c). Turn them around a vertical axis passing through the torso, while the shoulders remain completely horizontal. Vice versa, move the right hand to outline a quarter turn backwards, while the left hand moves forward. Repeat once, twice, or three times. In each case, gently allow the pelvis to lightly rise and descend.

Each asymmetrical movement, depending on whether it is rising and descending, front and back, or from side to side, addresses different areas of the abdominal region. When clearly perceived internally, the exercises actively combine imagination and shear forces, thus providing a gentle organ

3b: Massaging the organs - horizontal

3c: Massaging the organs - turning

4a: Circular movement . . .

4b: . . . carry and embrace

massage and the release of obstructions. To conclude, allow the hands to come back to stillness, ending on the same level at the hips.

5. Circular movement: Carrying and Enbracing

Open the hands, palms down, and move them in a circle away from the center of the body, gradually letting them rise to chest level (#4a; rising). Then, arching them to the sides (#4b), let them again descend to complete the circle in front of the abdomen (descending). In the process turn the palms over, so they end facing up (#5a). As introduced here, "rising" and "descending" in the following means rising and descending of the pelvis.

After concluding this preparation, you can go onto either the first or second cycle.

5a: Balancing the breath, . . .

5b: . . . calming the mind

5c

5d

First Cycle, "Origin"

1. Balancing the Breath, Calming the Mind

After concluding the circular movement, stand with your feet shoulder-width apart, holding your hands in front of the abdomen, palms up (#5a). Allow your attention to sink down into your feet and roots, while raising the hands—about fist-width apart—in a slow and controlled movement, as if lifting a ball (rising).

Once at shoulder level, relax the wrists and turn the palms down, as if gently placing a ball on the water (#5b). Imagining a slight resistance, press it down more deeply (sinking; #5c). Once at abdomen level, relax the wrists again and allow the palms to turn up (#5d). Repeat three to five times. If you have high blood pressure, make sure not to raise the hands above chest level.

6a: Dividing the clouds, . . .

6b

6c: . . . lowering the moon

6d

2. Dividing the Clouds, Lowering the Moon

Guide the hands to the sides of the body, fingertips pointing obliquely to the ground, palms facing forward (#6a). As you think of elastic bands, allow your hands and feet to connect with some tension. Your feet anchored firmly in the ground, lift your hands with arms extended to about chin level as if dividing the clouds—think of lifting fog or releasing bad temper (rising; #6b).

Relax your elbows and wrists, allowing the fingers to turn forward and up from the palms. While still at chin level, move your hands toward each other, beginning with the pinkies, as if you were holding the moon (#6c). Lower it along the center line of the torso all the way to the elixir field (descending, #6d). From the abdomen, move the hands back to the sides of the body. Repeat three to five times.

As before (#4a-b), execute a circular movement, lifting the hands inward and lowering them in an outward direction, while at the same time pulling in the left foot. End with closed feet and the palms facing up in front of the abdomen.

7a: Looking back. . . .

7b

7c: . . . and leaving behind

7d

3. Looking Back and Leaving Behind

Lift your hands to shoulder level (rising). Relax the wrists and turn the palms down as if placing a ball on water. Move the left foot sideways out to shoulder width (#7a). Lower the hands with the ball to the level of the navel while sitting down more deeply on the cloud (descending; #7b).

From here, let the hands press two balls of air next to the hips as though you wanted to use them for support. Point the fingertips toward the center of the body while turning your head to the left (rising; #7c). Gaze into the distance and leave all things behind.

For the closing part of the movement, turn the head back to the center and move the hands toward each other in front of the center of the body (descending; #7d). Move the hands once again in a circular motion, lifting them inward to about chest level (rising), then lowering them outward back to the abdomen (descending). Pull in the left foot and end with palms up. Repeat the exercise on the other side, turning the head to the right. Repeat once or twice on each side.

To transition to the next exercise, once again move the arms in a circular motion to the abdomen and step the left foot into a parallel stance.[2]

---

[2] The principles that underlie these pratices are also the main building blocks of a more extensive qigong practice that Cong Yongchun 从永春 from Fuzhou 福州 taught some years ago (Bamberger-Hüfner 1997, 11). I learned this from my first teacher Jiang Wuche 江武彻 from Fuzhou in 1988 in Beijing as an excercise for qi absorption. He emphasized that the practice is best undertaken in the open air, preferably under a tree.

8a: Reverting to original light

8b: Carry the moon in your hands

8c: Turn the pearl of the elixir field

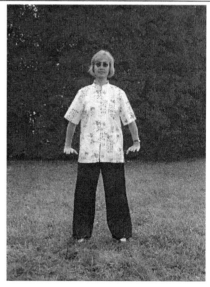

8d: Position of forceful warrior

4. Reverting to Original Light

Stand with feet shoulder-width apart. Place your hands on top of each other over the elixir field, so that the thumb of the lower hand is close to the navel (#8a). Collect your breath and attention there. Release the thumbs from the belly so that the hands can turn to face palms up: *carry the moon in your hands*. Pull the hands apart to the point where the fingertips no longer touch but still point toward each other (#8b).

Next, to *turn the pearl of the elixir field*, relax the wrists and turn the hands palms down, allowing the hands to rise to the level of the navel (rising; #8c). Lower the hands to the center of the abdomen (descending), open them to come next to the hips into the *position of forceful warrior* (rising; #8d).

8e: Pulling strings out of the soil

8f: Hold the sun and the moon in your hands

8g: Join the sun and the moon

8h: Return the qi to the origin

Continue to gaze straight ahead. Release the hands and direct both your fingers and your attention to the ground (descending). Extend the arms and lift them sideways, imagining that you are *pulling strings out of the soil* (rising; #8e). Once the arms reach shoulder level, lower the elbows and turn the palms up: *hold the sun and the moon in your hands* (#8f).

Bring the hands across your forehead to hold a large ball of air: *join the sun and the moon* (#8g). Lowering your hands, internally guide the ball through the center of your body along the Penetrating Vessel (Chongmai 沖脉) (descending): *return the* qi *to the origin* (#8h). End with both hands on top of each other over the elixir field.

Repeat three to six times, always very slowly and in deep inner stillness. The practice enacts yin and yang returning to Dao. Emerging from the center, the polarities divide to form Heaven and Earth, with the sun and the moon on the outside. Guided by imagination, they come back together within the individual human being.

9a: Pushing the mountain

9b

10a: Touching the feet . . .

10b

Second Cycle, "Transformations"

1. Pushing the Mountain

After readying yourself with the five preparation exercises, pull in the left foot while performing a circular movement with your arms. Roll a ball with your hands in front of the abdomen to end with palms facing out and down, thumbs and index fingers joined in a triangle and forming the "double tiger's mouth."

Step the left foot out into a bow stance, setting the foot down heel first, in an angle of about 45 to 60 degrees and about two feet distant from the right foot.[3] As you shift weight, move the hands forward to about shoulder height, as if you were pushing a mountain (#9a). The strength needed for this comes from the roots; the pelvis descends slightly.

Relax the elbows and wrists so the palms can turn inward and up, beginning with the pinkies. Imagine you are letting silk threads pass through your fingers (#9b). Slowly pull the hands back to abdomen level and at the same time bring the left foot back to a closed stance (rising). Turn the hands back to their original position, palms facing front and down.

Repeat to the right. Do once or twice more on each side. To transit to the next exercise, move your arms in a circle from inside to outside.

---

[3] For a more detailed description, see Bock-Möbius (1993, 46).

10c

10d: . . . and strengthening the hips

10e

10f

2. Touching the Feet and Strengthening the Hips

Hold your hands in front of the elixir field, palms facing up. Send your intention into the soles of your feet to activate a deep rooting in the ground and initiate the impulse to lift the ball to shoulder height (rising). Once there, turn it and place it on the water and step the left foot out to shoulder width (#10a).

Sit down more deeply, lowering the hands with the ball to the level of the navel (descending). Straighten up and guide the hands gently to the sides of the hips (rising). Move them back slowly, as if pulling them through water (descending; #10b). Allow them to come forward again, keeping the palms facing back, then continue the movement to lift the hands over the head (rising). Turn the palms up as if supporting the sky (#10c).

Keeping both the torso and the legs straight, bend forward and touch your feet (#10d). If you have high blood pressure, do not bend beyond ninety degrees. Keeping the torso horizontal, mentally connect your palms with the soles of your feet. Moving from the hips and keeping the legs and back straight, return to a standing position (#10e) while lowering the arms sideways to their original position and stepping the left foot back into a closed stance (#10f). Repeat on the right. Do once or twice more on both sides.

11a: Tiger fighting

11b

11c

11d

3. Tiger Fighting

Holding both hands in front of the elixir field, turn them palms down and mentally absorb balls of air into the Laogong 劳宫 points in the center of the palms. To do so, apply tiger claws—a force that collects and at the same time spans with some tension—between the joints at the base of thumbs, index fingers, and pinkies.[4] Shift your weight on to the right (#11a). Step your left foot into a bow stance (#11b), then move into the horse stance (feet wide and parallel) and from there into the tiger stance—by turning the left heel and the tip of the right foot slightly out, thus creating a lightly open foot position.

Relax the hands and use them to grip forward, as if picking something up from the floor, sitting down a bit more deeply at the same time (descending; #11c). Lift the object upward from the center of the body (rising). Once you reach head level, relax the wrists and turn both hands inward in a spiraling movement. This way the tiger claws point forward as if set to attack. At the same time, descend more deeply into the tiger stance (sinking; #11d). With the left hand protect your head, while using the right hand to shield chest and heart.

To release the position, shift your weight to the left (rising) and allow the hands to come closer toward each other in front of the torso. Bring the right foot in and lower your arms in a circular movement (descending). Repeat the exercise to the right. Do once or twice more on each side. The exercise is progressive: you move forward with every repetition. Aspects of the last three movements recur in the next exercise.

---

[4] For details, see Jiao (1992, 130).

12a: The five wheels

12b: Water

12c: Wood

12d

4. The Five Wheels

The five wheels embody the principle of change as it proceeds and re-volves through the five phases. They are also called "meditative swaying," since each phase involves an opening move with a backward (yin) and a closing move with a forward weight shift (yang).

The first phase is **water**. Beginning with a closed position, arms hang-ing loosely by the sides, step the left foot into a forward bow stance. Shift your weight onto the front foot, at the same time moving the hands on the sides of the body forward, palms in, as if you were pushing them through water (#12b). Relax the wrists and turn the palms forward. Next, shift your weight back, at the same time moving the hands backward, in your imagi-nation feeling water pass through fingers and over the back of the hand (#12a). Turn the palms back and once again repeat the forward movement while shifting the weight forward once more (#12b).

As you shift your weight once more toward the back, place the hands on the abdomen in the position of the double tiger's mouth (#12c). As you next shift forward, like the root of a plant penetrating the soil (representa-tive of the phase **wood**), bring your hands, fingertips raised upward, to face level (#12d). The pinkies touch; elbows are lightly bent.[5]

---

[5] Some versions of this practice require several repetitions of each phase, thus staying for some time with each, but in the way I learned it (Geitner 2003), each phase follows immediately from the preceding one. See also Kollak (2008, 48-62).

12e: Fire

12f

12g: Earth

12h

Shift your weight back again while spreading the arms to both sides and slightly back at chest level (#12e). Bring them back into the "tiger's mouth" position in front of the chest and, as you shift weight again, let them push forward (#12f). The first part of this movement shows an opening to, and acceptance of, the outside world; the second matches a return to, or engagement, in it. This represents the phase **fire**.

Lower the hands, as if to pick up something from the ground, to gather the fruits of the **earth**. Bring the right foot in and cross the hands in front of the belly (#12g). Step into a bow stance to the right and shift your weight forward as you turn the hands once more into the double tiger mouth position. Push them forward as if they were bringing their own fruits into the world (#12h).

12i: Metal

12j

12k: Transition to phase water

12l: Conclude with phase earth

Relax the wrists and turn the palms inward. At chest level, in line with the lungs and thus the phase **metal**, move the hands in a horizontal circle to the left (#12i). Imagine that you are embracing the rhythms of nature and giving your own rhythms to the world outside. Again, as you shift your weight forward, push the hands forward in the double tiger mouth position (#12j).

Lower the hands, passing the sides of the body as if pulling them through water, transiting once more to the phase **water** and shifting your weight backward (#12k). From here the cycle begins again. Repeat the entire sequence several times, moving each time during the phase **earth** by stepping into the bow stance either forward or back.

The practice ends with the phase **earth**. Cross the arms and step forward into closed stance (#12l), then lower the hands to rest on the sides of the body.

13: Rubbing the kidney points

14: Closing the belt vessel

15: Massaging the elixir field

16: Washing the Laogong points

Closure

1. Rubbing the Kidney Points

Stand with feet shoulder-width apart. Slowly rub your hands along the Bladder meridian, moving downward next to the lumbar spine and the kidney connection (*shenshu* 肾俞) points (#13). Then rub along the Governing Vessel (*dumai* 督脉), moving upward. Repeat this four times. Sit down a bit more deeply in the cloud as you push downward (descending); lift up somewhat as you rub upward (rising).

2. Closing the Belt Vessel

Use the ball of your hand to rub outward and forward from the spine along the Belt Vessel (*daimai* 带脉), which circles the body at the waist (#14). End with both hands on top of each other at the elixir field (#15).

3. Massaging the Elixir Field

Massage the elixir field in gradually enlarging circles. Keeping your hands on the belly, move them first right, then down, then left, and finally upward. Repeat four times, then change direction and let the circles become gradually smaller until the hands are back at the elixir field. As you do so, let the pelvis rise and descend gently. When executed in this manner, the exercise stimulates digestion.

4. Washing the Laogong Points

Place your hands in front of the upper torso, steepling them into a roof (#16). Lower the left hand as you raise the right, and vice versa. Repeat once. The descending movements bring the *qi* from the hands and arms back into the elixir field. Turn the palms toward the face and use the remaining *qi* to wash the face: stroke them lightly in front of the face, moving up and down. Again, let the pelvis rise and descend gently.

5. Returning the *Qi* to the Origin

Lift the hands all the way to eye level, then lower them to the sides. Bring the left foot in. This concluding circle lets the *qi* flow back into the elixir field. It allows you to leave the concentrated practice state and return to everyday consciousness.

Bibliography

Aristoteles. 2007. *Metaphysik*. Translated by Herman Bonitz. Reinbek bei Hamburg: Rowohlt.

Aspect, A., P. Grangier, and G. Roger. 1981. "Experimental tests of realistic local theories via Bell's theorem." *Physical Review Letters* 47: 460-63.

_____. 1982a. "Experimental realization of Einstein-Podolsky-Rosen-Bohm Gedanken-experiment: A new violation of Bell's inequalities." *Physical Review Letters* 49: 91-94.

_____, J. Dalibard, and G. Roger. 1982b. "Experimental test of Bell's inequalities using time-varying analyzers." *Physical Review Letters* 49: 1804-07.

Audretsch, Jürgen. 2002. *Verschränkte Welt: Faszination der Quanten*. Weinheim: Wiley-Vch.

Bamberger-Hüfner, Beate. 1997. "Die Übung vom 'Ursprünglichen Licht'." *Tiandiren Journal: Nachrichten der Deutschen Qigong Gesellschaft* 1: 11-12.

Bartl, Marlies. 2003. "Lü Dongin und die Acht Unsterblichen – die Entstehung und Entwicklung einer Legende." *Zeitschrift für Qigong Yangsheng* 2003: 67-77.

Bell, John S. 1964. "On the Einstein-Podolsky-Rosen paradox." *Physics* 1: 195-200. Reprinted in *Speakable and Unspeakable in Quantum Mechanics*. Cambridge: Cambridge University Press, 1987.

Belzig, W., and E. Scheer. 2008. "Physik IV: Integrierter Kurs; Atom- und Quantenphysik." Universität Konstanz Sommersemester.

Bennett, C.H., G. Brassard, and A. K. Ekert. 1992. "Quantum cryptography." *Scientific American* 267 (October): 50-57.

Bingen, Hildegard von. 2001. *Gott schauen*. Edited by H. Schipperges. Düsseldorf: Patmos.

Blofeld, John. 1985. *Das Geheime und Erhabene: Mysterien und Magie des Taoismus*. Munich: Goldmann.

Bock-Möbius, Imke. 1993. *Qigong – Meditation in Bewegung*. Heidelberg: Haug.

_____. 1996. "Untersuchungen des Qigong-Zustands mit Hilfe von EEG-Messungen." Bonn: Abschlussarbeit, Grundausbildung Medizinische Gesellschaft für Qigong Yangsheng.

_____. 2009. "Qigong oder das Quantum der Einheit. Wie Qigong Quantenphysik und Mystik versöhnt." *Zeitschrift für Qigong Yangsheng* 2009: 66-70.

_____. 2010a. "Qigong oder das Quantum der Einheit. Wie Qigong Quantenphysik und Mystik versöhnt." *Netzwerkmagazin* 2010: 8-11.

_____. 2010b. "Einheit erfahren. Daoismus und Ganzheit in der modernen Physik." *Taijiquan & Qigong Journal* 2: 46-50 und 3: 34-38.

_____. 2010c. "About the Relation between Qigong and Modern Physics." *Deutsche Zeitschrift für Akupunktur* 4: 46-51.

_____. 2011. "Daoism and entirety in quantum physics." *Journal of Daoist Studies* 4: 162-74.

Boente, Dorothea. 1999. "Qigong und Psychotherapie." *Zeitschrift für Qigong Yangsheng* 1999: 91-96.

Bohm, D., Aharonov, Y. 1957. "Discussion of experimental proof for the paradox of Einstein, Rosen, and Podolsky." *Physical Review* 108: 1070-76.

Bohm, David. 1952a. "A suggested interpretation of the quantum theory in terms of 'hidden' variables, Part I." *Physical Review* 85: 166-79.

_____. 1952b. "A suggested interpretation of the quantum theory in terms of 'hidden' variables,Part II." *Physical Review* 85: 180-93.

Bohr, Niels. 1935. "Can quantum mechanical description of physical reality be considered complete?" *Physical Review* 48: 696-702.

Borchert, Bruno. 1997. *Mystik: Das Phänomen - Die Geschichte - Neue Wege*. Freiburg: Herder.

Born, M., Heisenberg, W., Jordan, P. 1926. "Zur Quantenmechanik II." *Zeitschrift für Physik* 35: 557–615.

Bouwmeester, D., J.-W. Pan, K. Mattle, M. Eibl, H. Weinfurter, and A. Zeilinger, A. 1997. "Experimental quantum teleportation." *Nature* 390: 575-79.

Capra, Fritjof. 1988. *Das Tao der Physik. Die Konvergenz von westlicher Wissenschaft und östlicher Philosophie.* Munich: Scherz.

Castaneda, Carlos. 1973 [1968]. *Die Lehren des Don Juan: Ein Yaqui-Weg des Wissens.* Frankfurt a.M.: Fischer.

Chan, Wing-tsit. 1963. *A Source Book in Chinese Philosophy.* Princeton: Princeton University Press.

Chu, Hui-lien, P. Wrede. 1999. "Ein Universalgelehrter und seine Gedanken zum Yangsheng – Anmerkungen zu Su Shi." *Zeitschrift für Qigong Yangsheng* 1999: 66-73.

Cobos-Schlicht, Carlos. 1997. "Die Bedeutung der Einheit von Mensch und Natur in den emei-Merkversen." *Zeitschrift für Qigong Yangsheng* 1997: 16-30.

_____. 1998. "Die Leib-Seele-Geist-Vorstellung in der chinesischen Gedankenwelt anhand des Begriffes shen." *Zeitschrift für Qigong Yangsheng* 1998: 21-30.

Dahmer, Manfred. 2007. *Lass die Bilder klingen. Gedichte aus dem Chinesischen.* Uelzen: Medizinisch Literarische Verlagsgesellschaft.

Darga, Martina. 1999. *Das alchemistische Buch von innerem Wesen und Lebensenergie: Xingming guizhi.* Munich: Diederichs.

_____. 2004a. "Das Xingming guizhi und die Idee der Umwandlung." *Zeitschrift für Qigong Yangsheng* 2004: 60-67.

_____. 2004b. "Das Zinnoberfeld – dantian." *Zeitschrift für Qigong Yangsheng* 2004: 90-93.

Dawydow, A.S. 1972 [1963]. *Quantenmechanik.* Berlin: Deutscher Verlag der Wissenschaften, VEB.

Demtröder, Wolfgang. 2005. *Experimentalphysik 3: Atome, Moleküle, Festkörper.* Berlin: Springer.

Despeux, Catherine. 2007. "Zur Geschichte der chinesischen Medizin Teil 3." *Zeitschrift für Qigong Yangsheng* 2007: 26-31.

Dethlefsen, Thorwald, and Rüdiger Dahlke. 1983. *Krankheit als Weg: Deutung und Bedeutung der Krankheitsbilder.* Munich: Bertelsmann.

Einstein, A., B. Podolsky, and N. Rosen. 1935. "Can quantum-mechanical description of physical reality be considered complete?" *Physical Review* 47: 777-80.

Engelhardt, Ute. 1987. *Die klassische Tradition der Qi-Übungen (Qigong).* Stuttgart: Franz Steiner.

d'Espagnat, Bernard. 1979. "The quantum theory and reality." *Scientific American* 241:158-81.

Everett, Hugh III. 1957. "'Relative state' formulation of quantum mechanics." *Reviews of Modern Physics* 29: 454-62.

Feynman, R. P., R. B. Leighton, and M. Sands. 1999. *Feynman Vorlesungen über Physik: Bd. III, Quantenmechanik.* Munich: Oldenbourg.

Fischer Lexikon. 1981. *Das neue Fischer Lexikon in Farbe.* Frankfurt a.M.: Fischer.

Fischer, Ernst Peter. 2002. *Die andere Bildung: Was man von den Naturwissenschaften wissen sollte.* Munich: Ullstein.

_____. 2004. *Eine Welt, die keine Teile hat: Niels Bohr oder die Lektion der Quanten.* Gelsenkirchen: Edition Archaea.

Forman, Robert K. C. 1990. *The Problem of Pure Consciousness.* Oxford: Oxford University Press.

Freedman, S.J., and J. F. Clauser. 1972. "Experimental test of local hidden-variable theories." *Physical Review Letters* 28: 938-41.

Friedrichs, Elisabeth. 2003. "Qigong-Yangsheng-Übungen in der Begleitbehandlung bei Migräne und Spannungskopfschmerz." *Zeitschrift für Qigong Yangsheng* 2003: 101-12.

Geißler, Manfred. 1996. "Qigong-Yangsheng-Aktivitäten in den Niederlanden." *Zeitschrift für Qigong Yangsheng* 1996: 80-85.

Geitner, Richard. 2003. *Vierte Radolfzeller Qigong-Tage.* Mettnau-Kur, Radolfzell.

Gernet, Jacques. 1988. *Die chinesische Welt.* Frankfurt a.M.: Suhrkamp.

Girardot, Norman. 2009 [1983]. *Myth and Meaning in Early Taoism: The Theme of Chaos (Hundun).* Dunedin, Fla.: Three Pines Press.

Goepper, Roger. 2004. "Mensch und Natur – Chinesische Landschaftsmalerei." *Zeitschrift für Qigong Yangsheng* 2004: 27-51.

Goethe, J. W., and E. Trunz. 1963. *Goethes Faust: Der Tragögie erster und zweiter Teil, Urfaust. Kommentiert von Erich Trunz.* Hamburg: Wegner.

Graham, A. C. 1986. *Yin-Yang and the Nature of Correlative Thinking.* Singapore: The Institute for East Asian Philosophies.

_____. 1990. "The Origins of the Legend of Lao Tan." In *Studies in Chinese Philosophy and Philosophical Literature*, edited by A. C. Graham, 111-24. Albany: State University of New York Press.

Greene, Brian. 2000. *Das elegante Universum: Superstrings, verborgene Dimensionen und die Suche nach der Weltformel.* Berlin: Siedler.

Haffelder, Günter. 2006. *Wirkungen von Qigongübungen auf das Gehirn.* Stuttgart: Institut für Kommunikation und Gehirnforschung.

Heisenberg, Werner. 1925. "Über quantentheoretische Umdeutung kinematischer und mechanischer Beziehungen." *Zeitschrift für Physik* 33: 879-93.

_____. 1969. *Der Teil und das Ganze: Gespräche im Umkreis der Atomphysik.* Frankfurt a.M.: Büchergilde Gutenberg.

_____. 1974. *Across the Frontiers.* New York: Harper and Row. Quoted according to Ricard 2008.

Hildenbrand, Gisela. 1993. "Qigong – Gesundheitsfördernde Übungen der traditionellen chinesischen Medizin." *Qigong Yangsheng* 1993: 6-11.

_____. 1996. "Das Lehrsystem Qigong Yangsheng von Jiao Guorui — Kurzer Abriss über ausgewählte Methoden und ihre inneren Verbindungen." *Zeitschrift für Qigong Yangsheng* 1996: 12-30.

_____. 2000. "Unterrichtsmaterialien von Jiao Guorui zum Lehrsystem Qigong Yangsheng." *Zeitschrift für Qigong Yangsheng* 2000: 5-15.

_____. 2001. "Das Lehrsystem Qigong Yangsheng – Inhalte und Eigenschaften." *Zeitschrift für Qigong Yangsheng* 2001: 6-13.

_____. 2007. Vertiefungskurs „Das Spiel der 5 Tiere." Augsburg, Dezember.

Hofmann, Iris. 1999. "Qigong Yangsheng in der Bewegungstherapie bei koronarer Herzerkrankung." *Zeitschrift für Qigong Yangsheng* 1999: 105-06.

Huxley, Aldous. 1946. *The Perennial Philosophy*. New York and London: Harper & Brothers.

Jäger, Willigis. 1991. *Suche nach dem Sinn des Lebens: Bewusstseinswandel durch den Weg nach innen*. Petersberg: Via Nova.

_____. 2000. *Die Welle ist das Meer: Mystische Spiritualität*. Freiburg: Herder.

_____. 2005. *Wiederkehr der Mystik: Das Ewige im Jetzt erfahren*. Freiburg: Herder.

_____, and Beatrice Grimm. 2004. *Der Himmel in dir. Einübung ins Körpergebet*. Munich: Kösel.

Jämlich, Manfred. 2000. "Qigong und Psychotherapie – westliche Perspektiven." *Zeitschrift für Qigong Yangsheng* 2000: 76-81.

Jiao, Guorui. 1988a. *Qigong Essentials for Health Promotion*. Beijing: China Reconstructs Press.

_____. 1988b. *Qigong Yangsheng: Gesundheitsfördernde Übungen der traditionellen chinesischen Medizin*. Uelzen: Medizinisch Literarische Verlagsgesellschaft.

_____. 1989. *Die 15 Ausdrucksformen des Taiji-Qigong: Gesundheitsfördernde Übungen der traditionellen chinesischen Medizin*. Uelzen: Medizinisch Literarische Verlagsgesellschaft.

_____. 1992. *Das Spiel der 5 Tiere: Qigong; gesundheitsfördernde Übungen der traditionellen chinesischen Medizin*. Uelzen: Medizinisch Literarische Verlagsgesellschaft.

_____. 1993a. *Qigong Yangsheng – Ein Lehrgedicht*. Uelzen: Medizinisch Literarische Verlagsgesellschaft.

_____. 1993b. "Die Bedeutung und Methodik des In-die-Ruhe-Tretens bei Qigong-Übungen." *Qigong Yangsheng* 1993: 12-15.

_____. 1994. "Die wichtige Funktion der Vorstellungskraft bei der Mobilisierung und Kontrolle der Lebenskraft." *Qigong Yangsheng* 1994: 7-12.

_____. 1996a. *Die 8 Brokatübungen: Bewegung und Ruhe. Qigong Yangsheng; gesundheitsfördernde Übungen der traditionellen chinesischen Medizin*. Uelzen: Medizinisch Literarische Verlagsgesellschaft.

_____. 1996b. "Der Kommentar zum Taijiquan (Taijiquan lun) von Wang Zongyue: Einführung, textkritische Anmerkungen und Übersetzung." *Zeitschrift für Qigong Yangsheng* 1996: 7-11.

_____. 1996c. "Imagination und Inspiration in der Übungspraxis des Qigong Yangsheng." *Zeitschrift für Qigong Yangsheng* 1996: 36-40.

_____. 1997a. "Abriss des Taiji jinggong 'Taiji-Übung-in-Ruhe'." *Zeitschrift für Qigong Yangsheng* 1997: 9-15.

_____. 1997b. "Zwölf Aspekte einer Vertiefung der Übungspraxis des Qigong Yangsheng." *Zeitschrift für Qigong Yangsheng* 1997: 75-79.

Kaptchuk, Ted J. 1994 [1983]. *Das große Buch der chinesischen Medizin: Die Medizin von Yin und Yang in Theorie und Praxis*. Munich: Heyne.

_____. 2000. *The Web that Has No Weaver: Understanding Chinese Medicine*. New York: Congdon & Weed.

Kohn, Livia. 1989a. "Guarding the One: Concentrative Meditation in Taoism." In *Taoist Meditation and Longevity Techniques*, edited by Livia Kohn, 123-56. Ann Arbor: University of Michigan, Center for Chinese Studies Publications.

_____. 1989b. "Taoist Insight Meditation: The Tang Practice of *Neiguan*." In *Taoist Meditation and Longevity Techniques*, edited by Livia Kohn, 191-222. Ann Arbor: University of Michigan, Center for Chinese Studies Publications.

_____. 1992. *Early Chinese Mysticism: Philosophy and Soteriology in the Taoist Tradition*. Princeton: Princeton University Press.

_____. 1997. "Yin and Yang: The Natural Dimension of Evil." In *Philosophies of Nature: The Human Dimension*, edited by Robert S. Cohen and Alfred I. Tauber, 89-104. New York: Kluwer Academic Publishers, Boston Studies in the Philosophy of Science.

_____. 1998. "Mind and Eyes: Sensory and Spiritual Experience in Taoist Mysticism." *Monumenta Serica* 46: 129-56.

_____. 2005. *Health and Long Life: The Chinese Way*. Cambridge, Mass.: Three Pines Press.

_____. 2008. *Chinese Healing Exercises: The Tradition of Daoyin*. Honolulu: University of Hawai'i Press.

_____. 2010. *Sitting in Oblivion: The Heart of Daoist Meditation*. Dunedin, Fla.: Three Pines Press.

_____, and Michael LaFargue, eds. 1998. *Lao-tzu and the Tao-te-ching*. Albany: State University of New York Press.

Kollak, Ingrid, ed. 2008. *Burnout und Stress: Anerkannte Verfahren zur Selbstpflege in Gesundheitsberufen.* Heidelberg: Springer.

Kopp, Clemens. 1997. "Wenn man nur beständig und gewissenhaft übt. Qigong und die philosophische Lehre vom Habitus." *Zeitschrift für Qigong Yangsheng* 1997: 99-101.

LaFargue, Michael. 1994. *Tao and Method: A Reasoned Approach to the Tao Te Ching.* Albany: State University of New York Press.

Laszlo, Ervin. 2005. *Zuhause im Universum: Eine neue Vision der Wirklichkeit.* Berlin: Allegria.

Liu, Tianjun. 2005. "Vorbereitung auf das 'In-die-Ruhe-Treten'." *Zeitschrift für Qigong Yangsheng* 2005: 90-94.

Liu, Xiaogan. 1998. "Naturalness (*Tzu-jan*), the Core Value in Taoism: Its Ancient Meaning and Its Significance Today." In *Lao-tzu and the Tao-te-ching*, edited by Livia Kohn and Michael LaFargue, 211-28. Albany: State University of New York Press.

_____. 2001. "Nonaction and the Environment Today: A Conceptual and Applied Study of Laozi's Philosophy." In *Daoism and Ecology: Ways Within a Cosmic Landscape*, edited by Norman Girardot, James Miller, and Liu Xiaogan, 315-40. Cambridge, Mass.: Harvard University Press, Center for the Study of World Religions.

Louis, Francois. 2003. "The Genesis of an Icon: The Taiji Diagram's Early History." *Harvard Journal of Asiatic Studies* 63: 145-96.

Needham, Joseph, et al. 1956. *Science and Civilisation in China*, vol. II: *History of Scientific Thought.* Cambridge: Cambridge University Press.

_____. 1976. *Science and Civilisation in China*, vol. V.3: *Spagyrical Discovery and Invention—Historical Survey, from Cinnabar Elixir to Synthetic Insulin.* Cambridge: Cambridge University Press.

Neuhaus, Elke. 2001. "Überlegungen zum Einsatz von Qigong Yangsheng im Rahmen bewegungstherapeutischer Verfahren bei chronischen psychiatrischen Störungen." *Zeitschrift für Qigong Yangsheng* 2001: 102-10.

Ommerborn, Wolfgang. 2003. "Das Studium alter Schriften als Teil des Erkenntnisprozesses im Neo-Konfuzianismus." *Zeitschrift für Qigong Yangsheng* 2003: 45-56.

_____. 2004a. "Der Neo-Konfuzianismus der Song-Zeit." *Zeitschrift für Qigong Yangsheng* 2004: 68-79.

_____. 2004b. "Der Begriff Ziran [Qi/Wu/Wuwei/Ren] in der chinesischen Geistesgeschichte." *Zeitschrift für Qigong Yangsheng* 2004: 100-18.

_____. 2005a. "Einführung in den Daoismus." *Zeitschrift für Qigong Yangsheng* 2005: 35-56.

_____. 2005b. " Der Begriff Li in der chinesischen Geistesgeschichte." *Zeitschrift für Qigong Yangsheng* 2005: 100-105.

_____. 2006. "Der Begriff Taiji [Das Begriffspaar Yin und Yang] in der chinesischen Geistesgeschichte." *Zeitschrift für Qigong Yangsheng* 2006: 17-20, 28-35.

_____. 2007a. "Einführung in den Buddhismus in China." *Zeitschrift für Qigong Yangsheng* 2007: 48-72.

_____. 2007b. "Der Begriff Chaos (hundun) in der chinesischen Geistesgeschichte." *Zeitschrift für Qigong Yangsheng* 2007: 86-91.

Palmer, David. 2007. *Qigong Fever: Body, Science and Utopia in China*. New York: Columbia University Press.

Planck, Max. 2008. Originalton in "Nano Extra: Max Planck – Die körnige Welt," 3Sat TV Show, April 9 (22:25).

Pohl, Karl-Heinz. 2007. "Symbolik und Ästhetik der chinesischen Bambusmalerei." *Zeitschrift für Qigong Yangsheng* 2007: 32-47.

Powers, Margaret Fishback. 1996. *Spuren im Sand: Ein Gedicht, das Millionen bewegt, und seine Geschichte*. Gießen: Brunnen.

Quint, Josef, ed. 2007 [1963]. *Meister Eckehart: Deutsche Predigten und Traktate*. Hamburg: Nikol.

Reerink, Gertrud. 1997. "Die Entdeckung abrufbarer Ruhe." *Zeitschrift für Qigong Yangsheng* 1997: 96-98.

Reuther, Ingrid. 1996. "Qigong Yangsheng in der Behandlung von Asthma." *Zeitschrift für Qigong Yangsheng* 1996: 44-50.

Ricard, M., and T. X. Thuan. 2008. *Quantum und Lotus: Vom Urknall zur Erleuchtung*. Munich: Goldmann.

Ritter, Christine. 2000. "Qigong Yangsheng als Zusatztherapie bei Bluthochdruck im Vergleich zu einer westlichen Entspannungstherapie." *Zeitschrift für Qigong Yangsheng* 2000: 82-87.

Robinet, Isabelle. 1989. "Original Contributions of *Neidan* to Taoism and Chinese Thought." In *Taoist Meditation and Longevity Techniques,* edited by Livia Kohn, 297-330. Ann Arbor: University of Michigan, Center for Chinese Studies Publications.

Roth, Harold D. 1999. *Original Tao: Inward Training and the Foundations of Taoist Mysticism.* New York: Columbia University Press.

Röthlein, Brigitte. 2004. *Die Quantenrevolution: Neue Nachrichten aus der Teilchenphysik.* Munich: Deutscher Taschenbuch Verlag.

Sandleben, W.-I., R. Schläpfer. 1997. "Die Wirkung von Qigong Yangsheng nach kurzer Übungspraxis." *Zeitschrift für Qigong Yangsheng* 1997: 108-18.

Schäfer, Lothar. 2004. *Versteckte Wirklichkeit. Wie uns die Quantenphysik zur Transzendenz führt.* Stuttgart: Hirzel.

Schild, Wolfgang. 1997. "Körperertüchtigung und Leibbemeisterung." *Zeitschrift für Qigong Yangsheng* 1997: 51-63.

Schrödinger, Erwin. 1926a. "Quantisierung als Eigenwertproblem (Zweite Mitteilung)." *Annalen der Physik* 79: 489-527.

_____. 1926b. "Über das Verhältnis der Heisenberg-Born-Jordanschen Quantenmechanik zu der meinen." *Annalen der Physik* 79: 734-56.

_____. 1935a. "Die gegenwärtige Situation in der Quantenmechanik. Teil I." *Die Naturwissenschaften* 23: 807-12.

_____. 1935b. "Die gegenwärtige Situation in der Quantenmechanik. Part II. *Die Naturwissenschaften* 23: 823-8.

_____. 1935c. "Die gegenwärtige Situation in der Quanten-mechanik. Part III. *Die Naturwissenschaften* 23: 844-9.

Schubert, Stephan. 2000. "Psychotherapeutische Wirksamkeitsfaktoren im Qigong Yangsheng." *Zeitschrift für Qigong Yangsheng* 2000: 65-70.

Schwarz, Ernst. 1985. *Laudse: Daudedsching.* Munich: Deutscher Taschenbuch Verlag.

Scully, M.O., B.-G. Englert, and H. Walther. 1991. "Quantum optical tests of complementarity." *Nature* 351: 111-16.

Shiatsu Journal: *Information der Gesellschaft für Shiatsu in Deutschland*. 2008/53: 47.

Störig, Hans Joachim. 1985. *Weltgeschichte der Philosophie*. Frankfurt a.M.: Büchergilde Gutenberg.

Swami Sudip. 2004-2008. Lectures at Summer Retreats. Near Munich.

Tan, Dajiang. 2006. "Daoismus und Taijiquan." *Zeitschrift für Qigong Yangsheng* 2006: 83-91.

Tegmark, M., and J. A. Wheeler. 2001. *Spektrum der Wissenschaft* April: 68-76.

Tittel, W., J. Brendel, B. Gisin, T. Herzog, H. Zbinden, and N. Gisin. 1998. "Experimental demonstration of quantum correlations over more than 10 km." *Physical Review A* 57: 3229-32.

Wheeler, J.A., and K. Ford. 1998. *Geons, Black Holes & Quantum Foam: A Life in Physics*. New York: Norton.

Wilber, Ken. 1990 [1980]. *Das Atman Projekt: Der Mensch in transpersonaler Sicht*. Paderborn: Junfermann.

_____. 1996 [1991]. *Mut und Gnade: In einer Krankheit zum Tode bewährt sich eine große Liebe*. Munich: Goldmann.

_____. 1998. *Naturwissenschaft und Religion: Die Versöhnung von Weisheit und Wissen*. Frankfurt a.M.: Krüger.

_____. 2004 [1981]. *Halbzeit der Evolution: Der Mensch auf dem Weg vom animalischen zum kosmischen Bewusstsein*. Frankfurt a.M.: Fischer.

Wilhelm, Hellmut. 1948. "Eine Chou-Inschrift über Atemtechnik." *Monumenta Serica* 13: 385–88.

Wilhelm, Richard. 1950. *The I Ching or Book of Changes*. Princeton: Princeton University Press, Bollingen Series XIX.

_____. 2006 [1969]. *Dschuang Dsi. Das wahre Buch vom südlichen Blütenland*. Kreuzlingen: Hugendubel.

Zeh, H.D. 1970. "On the interpretation of measurement in quantum theory." *Foundations of Physics* 1: 69-76. Reprinted in: Wheeler, J.A., Zurek, W.H. 1983. *Quantum Theory and Measurement*. Princeton: Princeton University Press.

Zeilinger, Anton. 2005. *Einsteins Schleier. Die neue Welt der Quantenphysik*. Munich: Goldmann.

Zumfelde-Hüneburg, Christa. 1994a. "Einfluss der Qigong-Übungsmethode Tuna auf Parameter der Kreislauf- und Atemfunktion." *Qigong Yangsheng* 1994: 13-19.

_____. 1994b. "Berichte von Tagungen, Kongressen und Projekten." *Qigong Yangsheng* 1994: 67-68.

Zurek, Wojciech H. 1991. "Decoherence and the transition from quantum to classical." *Physics Today* 44: 36-44.

Index

CPSIA information can be obtained at www.ICGtesting.com
Printed in the USA
BVOW080556110512

289639BV00007B/1/P

9 781931 483216